COLOR-RICH
GARDENING
FOR THE SOUTH

A Guide for All Seasons

COLOR-RICH GARDENING
FOR THE SOUTH

ROXANN WARD

Original watercolors by Emmie Ruth Wise

THE UNIVERSITY OF NORTH CAROLINA PRESS   CHAPEL HILL

Publication of this book was supported in part by a generous gift from Florence and James Peacock.

Designed by Kim Bryant

Set in Arno by Rebecca Evans

Manufactured in the United States of America

The University of North Carolina Press has been a member of the Green Press Initiative since 2003.

All cover and interior photographs by Roxann Ward.

Library of Congress Cataloging-in-Publication Data

Names: Ward, Roxann, author. | Wise, Emmie Ruth, illustrator.

Title: Color-rich gardening for the South: a guide for all seasons / Roxann Ward; original watercolors by Emmie Ruth Wise.

Description: Chapel Hill: The University of North Carolina Press, 2021. | Includes bibliographical references and index.

Identifiers: LCCN 2020024186 | ISBN 9781469661766 (pbk. : alk. paper) | ISBN 9781469661773 (ebook)

Subjects: LCSH: Color in gardening—Southern States. | Gardening—Southern States—Handbooks, manuals, etc.

Classification: LCC SB454.3.C64 W37 2021 | DDC 635.0975—dc23

LC record available at https://lccn.loc.gov/2020024186

CONTENTS

FIGURES & TABLE

FIGURES

TABLE

PREFACE

My first attempt at gardening began twenty-some years ago, when a box arrived containing what gardeners refer to as "pass-along plants" (those plants easily divided and shared with others). These bits of root, lovingly wrapped in newspaper, did not hold much promise, and after some thought as to which end should go into the ground, I planted them in the loamy soil and gave them a drink from the garden hose. Wanda, a dear family friend, and my grandmother, Almeda, each sent me a box that summer—one containing starts of bearded iris, the other daylilies. These wise gardeners knew that a few foolproof perennials were the training wheels I needed to begin my gardening journey.

Two years later, after a bit of success with my perennials, I saw no reason why I should not have a similar experience with my new Georgia garden, which, I proudly noted, was within U.S. Department of Agriculture plant hardiness zone 7b, a true *southern* garden. I imagined beds of rich purple pansies, glowing daffodils, and billowing azaleas. That first year, it all went terribly wrong. It seemed that I could not make anything grow in the red Georgia clay.

Humbled, I spent the next several months pouring through every southern gardening book I could find at my local library. I tried to understand exactly what it would take to make the garden in my imagination a reality. In spring, with garden books in hand, I began again.

After gardening in Georgia for several years, my family moved to England. For two years, I wandered along village paths where pale-lemon corydalis sprouted from cracks in ancient rock walls in summer, and snowdrops dotted the ground in late winter. I was living every American gardener's dream. During my last summer in England, I studied garden design at

Figure P.1. Surprise lilies (*Lycoris radiata*) bloom on bare stems in early fall.

a local agricultural college, and later, upon returning home, I returned to college and began a second career as a seasonal color designer in Atlanta.

It was there that I learned to create color-filled garden spaces for customers, through every season, using a wide variety of plant material. Over the years, I've learned to use combinations of ornamental shrubs, vines, tropical foliage plants, perennials, and annuals in my designs. I've grown to appreciate the many attributes of traditional southern plants, such as *Camellia sasanqua* and *Kalmia latifolia*. I hope to share these ideas with you, as you create your own unique color displays using the best of southern plants, old and new.

ACKNOWLEDGMENTS

I could not have written this book without the support and encouragement of my husband, Mark, who doesn't understand my endless need to grow things, but loves me anyway.

My children, Katherine, Nicholas, and Ana, will always be the best part of my gardening memories, and I'm immensely proud of each of them. Their childhood flower drawings often float down from the pages when I open my old gardening books, a reminder of sunny days long ago. I am thankful for all of the times they cheered me on during this process.

I want to thank my college friend Marni Jameson, an accomplished writer, author, and businesswoman, who assured me that I was not crazy when I asked for her advice about a gardening book I had begun to write. She gave me the short course on what to expect when working with a publisher, and for that I'm very grateful.

Thank you to Emmie Ruth Wise, the talented artist who produced the beautiful watercolor illustrations for chapter 5 of this book. She approached this project with curiosity, imagination, technical skill, and discipline, and I am honored to include her work in my book.

Finally, I want to thank my editor, Elaine Maisner, at the University of North Carolina Press, for her encouragement as I wrote and edited this manuscript. Thank you to the staff members who answered my questions and guided me through this process. I'm honored to have my book included with the titles published by the oldest university press in the South.

COLOR-RICH
GARDENING
FOR THE SOUTH

INTRODUCTION
THE ROOTS OF SOUTHERN GARDENING

In 1786, a Frenchman named André Michaux came to Charleston, South Carolina, where he eventually established a hundred-acre garden just outside the city. His botanical expeditions were fruitful, and Michaux is said to be responsible for bringing many old-world species to both France and America, including the tea olive, camellia, and crape myrtle. He is also credited with the discovery of hundreds of plants native to the Carolinas, including the sunflower, blazing star, flame azalea, bigleaf magnolia, and catawba rhododendron. His love for botany left a lasting mark on the southern landscape.

Figure I.1. Coleus, ferns, and begonias bring texture and color to a shady garden area.

A hundred years later, during the Victorian era, gardeners planted showy flowering shrubs, such as camellias and azaleas, along with beds of dianthus, impatiens, periwinkles, and pansies to create beautiful displays. Today, in our southern gardens, you'll find these beloved flowering shrubs combined with new cultivars of the many annuals, perennials, and bulbs found in Victorian gardens. When we include these colorful plants in our designs, they not only give a sense of place to our gardens, they also connect us to the past in the best possible way. They help us remember the grandparents and great-grandparents who gardened before us.

Creating Modern Southern Gardens

Today one of the best examples of the southern gardening spirit can be found in the window boxes of Charleston. These botanical jewels charm us no matter the season. Through the heat of summer and the chill of winter, Charlestonians fill their window boxes with all manner of plant material: flowers, ferns, vines, bulbs, woody ornamentals. This southern gem, and others like it, have inspired gardeners for generations.

How can I help you bring the southern gardening spirit into your own garden? First, I'll explain how to pinpoint the best places to use flowering plants and teach you how to prepare planting areas. Then, I'll guide you through the process of planning and designing a color display, using a wide variety of plant material, from shrubs and flowers, to herbs and greens. Notes are included with each example, so you'll understand the "why" and "how" behind each design. I'll delve into the topic of garden maintenance to provide you with common sense advice on everything from pruning to pest control. Throughout the book, I'll share the shortcuts and strategies I've learned over the years to help you, wherever you find yourself on your gardening journey.

Figure I.2. Kimberly queen ferns, caladiums, pentas, and dragon wing begonias thrive in light shade.

Southern Planting Zones

Designs in this book include plant material appropriate for U.S. Department of Agriculture plant hardiness zones 6b through 8b, regions where mild winters allow gardeners to create showy flower displays year-round. This book has been written for gardeners in the following states and regions: Alabama, Arkansas, Delaware, the District of Columbia, Georgia, Kentucky, Louisiana, Maryland, Mississippi, North Carolina, North Central Texas, Northern Florida, South Carolina, Southern Missouri, Southern Oklahoma, Tennessee, Virginia, and West Virginia. While this book has been written primarily for southern gardens, the designs I share can easily be adapted to any area of the country.

Figure I.3. Cape plumbagos, zinnias, lantanas, and fan flowers create layers of color in a sunny border.

{ 1 }

SELECTING A SITE FOR YOUR COLOR GARDEN PROJECT

The best way to plan any garden space is to begin with a series of questions, and those questions are best answered *before* you find yourself in the aisle of a garden center, weak in the knees, bewitched by a stunning peony in full bloom. If you watch garden shoppers anywhere in the South on a beautiful spring day, you will conclude that it takes a great deal of willpower not to walk out of the store with an armful of plants too pretty to pass up. You and I have both done it. This is what I call gardening backward. It happens when you fill a shopping cart with flowers, confident that all will be beautifully arranged, once you figure out where they should go. Instead, I would challenge you to start at the beginning.

Figure 1.1. Coleus, dragon wing begonias, pentas, caladiums, lantanas, and fan flowers create a colorful summer display.

Planning before Design

Every garden area, large or small, should have a purpose. Flowers might highlight the front porch of your home, where you greet family and friends, or bring life to a shady spot where you like to relax and unwind in the evening. Think about how you use your outdoor spaces. Are you entertaining, playing with your children, or reading the paper?

Here are a few questions to help you get started:

- Where do you like to entertain? If you enjoy warm weather gatherings with friends and family, is there space for a flower bed or container grouping to create a bit of drama?
- Would you like to add color to the front entrance of your home? If so, what would you like the plants in this space to look like? Do you envision a pair of elegant, evergreen shrubs in containers, surrounded by lush flowers? Or flower beds along the sidewalk, with layers of colorful plants?
- Perhaps container gardening fits best into your busy life. Is there a sunny spot on a terrace or porch where you might add a group of containers to hold seasonal flowers and herbs?
- Would you like to add colorful native plants to your garden to attract birds or butterflies?
- If your goal is to grow flowering plants in a meandering sunny border, what might it look like in midsummer? Is it filled with flowering shrubs? Or is it a mixed planting with shrubs, perennials, and annuals?
- Growing your own vegetables and fruit may be something you'd like to try on a small scale. Have you considered a small planting bed for blueberries in summer, and kale and lettuce in fall?

Once you've considered all the places you might use seasonal color, choose one for your first project, knowing you can tackle others when time and budget allow. Perhaps it's the back entrance to your home where

cheerful containers of flowers might greet you at the end of a long day. It could be an out-of-the-way spot fitted with an Adirondack chair, where you sip coffee on the weekends surrounded by fragrant gardenias and summer flowers. In the end, seasonal color is not about impressing the neighbors with grand flower beds, it's about creating a joyful place that will draw you in and welcome you home.

Figure 1.2. Lamb's ears, black-eyed Susans, pentas, angelonias, and fan flowers bring life to a city garden.

{ 2 }

SITE ANALYSIS AND PLANNING

Once you have an idea in mind for your color project,
a bit more planning is in order before you get out your garden
tools. Let's consider the basics of location, light, water,
and soil.

Does the space you've chosen for your project *truly* meet your goals?

If your goal is to have flowers blooming near your front porch to welcome visitors, don't add a flower bed around your mailbox simply because the neighbor across the way has always done so, unless of course, you wish to impress your mail carrier. Make the most of your plant budget by having a game plan and sticking to it.

If you have one or two existing flower beds, evaluate them with a critical eye. Do they achieve a goal that is in line with your new color plan? If the flower bed in question was added on a whim, is there a better place to use color in your landscape? This is especially important when you have a budget in mind for your project. When you spend money on plant

Figure 2.1. Bellflowers brighten a garden in part shade.

material, make it count. One sizeable flower bed shows up better in the landscape than five tiny flower beds with a few flowers in each one.

The best way to achieve a design goal may not be a flower bed at all—it may be a large container. Containers are an easy way to add color through the seasons, and since they are portable, they can be rearranged whenever you wish to change the display.

Are light conditions in the proposed area appropriate for the plants you wish to grow?

If the proposed area receives sun all day, and you have sun-loving plant material in mind for your project, there will be plenty of light to grow anything you wish. If the area is shaded for part of the day, or is in the filtered shade of tree branches, you will need to determine how much sun and how much shade it receives so that you can choose appropriate plant material for your project.

A garden area in part shade is one that receives two to six hours of bright sunlight during the day. The light shade cast by high tree branches, such as pines, is helpful because it shields sun-loving plants from the harshest rays of the sun in summer. Full shade areas receive little sunlight during the day and are perfect for plants, such as ferns, found in shady woodlands.

Is there easy access to water?

Be sure you'll be able to water plants easily when nature doesn't provide rainfall. If you don't have an irrigation system for your property, there are timers and drip systems available to make the task of watering easier. In most cases, a simple garden hose and variable spray nozzle are all you need to keep your plants watered.

Does the existing soil drain well?

It's important to choose an area with good drainage for your garden project. Perform this simple test to be sure there are no problems with the area:

dig a hole several inches deep and fill it with water. If the water does not drain away within twenty-four hours, choose another area for your project. Drainage issues can be addressed by a landscaping professional. The answer to your drainage problem may be as simple as redirecting water from downspouts, or as complex as installing a French drain to carry away excess water from your yard.

While well-draining soil is ideal for most gardening projects, all is not lost if you have areas with damp soil. Do a little research, and you'll find many plants that don't mind wet feet, including natives such as bee balm, cardinal flower, and turtlehead.

Have you decided on a size for planting beds or containers?

Once you pinpoint an area for your project, you will need to decide on the size of the bed or container. The designs I've included in chapter 5 can easily be adapted to your garden, so if you only have space for a small raised bed where you will grow flowers and vegetables, pick and choose what works for you from a given design. This may mean your vegetable garden has a single large rosemary to anchor the bed, with vegetables or greens planted around it, and an edging of low-growing flowers for color. If you are new to gardening, it will be easier to start with a small space than to plan a 100-square-foot border, which might seem overwhelming by midsummer. A relatively small space can be packed with color if you plan well and make changes along with the seasons.

A color display should be large enough to show up well in your landscape, based on the size of your home's footprint. To help visualize a potential planting bed, sprinkle flour in a line around the perimeter of the proposed area to see if you like the shape and size. If the bed will contain a variety of plants, a good rule of thumb is to allow a *minimum* of four feet of depth for your planting bed. The deeper the bed, the more room you will have to experiment with small shrubs, perennials, annuals, and bulbs.

Beds containing shrubs should be deep enough to allow for growth over time if, for example, hydrangeas or other flowering shrubs are to remain in place for a few years (or longer). Seasonal color displays are meant to be

flexible and suit the whims of the gardener, with some components staying in place, while others are changed out according to the season. The designs in this book include ornamental shrub varieties that will grow to (or can easily be maintained at) a width of approximately four feet. This allows plenty of space for other colorful plant material, even in a small area. It is sometimes helpful to place bamboo stakes in the ground to mark potential shrub locations and check the spacing of shrubs to determine if you need to enlarge or shrink an area to accommodate the plants you wish to use.

If you want container plantings to make a big statement in your landscape, the containers you choose should be large enough to hold the variety of plants you wish to grow, but not so large that they look out of place. A good container size for most projects is somewhere between twenty-two and thirty-six inches in diameter. Be sure the container you choose has a hole in the bottom to allow for drainage.

Is there root competition from nearby mature trees or shrubs?

Large, established trees and shrubs have equally large root structures underground. In the South, annuals and perennials are less stressed in summer when their roots are not competing for moisture and nutrients with large root systems nearby. Use drought-tolerant groundcovers, such as barrenwort or pachysandra to dress up areas in dry shade near large trees and shrubs.

Figure 2.2. Gold edge durantas, dragon wing begonias, crossandras, angelonias, and lantanas thrive in summer heat.

{ 3 }
BED PREPARATION

If there is one thing you can do to give plants a good start, it is this: amend your soil with organic matter to improve drainage. While most trees and shrubs will grow happily in native soil, flowers and vegetables are more particular. Whether your project involves a bed of flowering shrubs and annuals, or a few vegetables and herbs with perennials, you will likely have more success if you amend the soil before you plant.

Soil Texture

Most of us have an idea whether our soil is clay or sand, or something in between. Depending on where you live, the soil on your land may vary from heavy clay, compacted during house construction, to the crumbly soil of an untouched woodland or grassy meadow. To know more about the soil in the specific area you've chosen for your garden project, perform the following simple test.

Figure 3.1. Violas bloom in a cool-season bed, while foxglove foliage provides a leafy background, promising height and drama in late spring.

You'll need a small, bare patch of soil, so use a shovel to remove any existing grass or plant material to reach the soil underneath. Dampen the area until it is just moist, take a handful of soil and try to form it into a ball. If it is slippery, it is mostly clay. If it feels gritty and falls apart, it has a high sand content. If it holds together loosely, it is closer to loam (which is a mixture of clay, silt, and sand). When you amend soil with organic matter, you improve its structural characteristics and ability to hold on to nutrients. If you are lucky enough to have loam soil that is friable and drains well, you may not need to add any organic matter before you begin your project.

Testing Your Soil

Before you begin any garden project, it's a good idea to have your soil tested. It is helpful to complete this step before you add amendments, and again afterward, so the test results will reflect any additional nutrients contained in the organic matter you've added to the bed. Order a soil testing kit on the Web, or contact your local county extension office to find out if they offer soil testing services in your area.

Follow the instructions closely when you collect your soil samples, and be sure to indicate on the test paperwork that you wish to grow flowers. This is important, because the recommendations you receive will be based specifically on the plant types you've listed (for example, turf, trees, shrubs, flowers, vegetables). After you've sent in your samples for testing, you'll receive a report showing levels of organic matter, pH, calcium, magnesium, sulfur, phosphates, potassium, sodium, chloride, boron, iron, manganese, copper, and zinc. Your test results will include recommendations on how to correct deficiencies (if any) before you plant.

Amending the Soil with Organic Matter

If you need to add organic matter to clay or sandy soil to improve its structure, you'll first need to remove any existing turf or weeds. To remove grass rhizomes, you may need to dig down two inches or more, or use a sod cutter to make the process easier.

Cover the entire planting area with one to three inches of organic matter or compost, working it in as deeply as you can with a spading fork. A rototiller is helpful for large areas. Composted manure, mushroom compost, well-composted leaves, or homemade compost are all good options.

Don't go overboard on soil amendments. Research shows that using more than a 20 to 30 percent volume of amendments can have a *negative* effect on plant growth. Once amendments are added, the bed will be a few inches higher than the surrounding soil. Rake the bed smooth, creating a flat top with gently sloping sides.

Allow for Prep Time

Ideally you should complete this process several weeks before you're ready to purchase plants for a spring or fall project. If you don't have time to prepare your soil for the upcoming season, try a small project first, such as a container planting. This will allow you to grow some of the flowering plants you've been eyeing at the garden center, but on a smaller scale, with no prep work involved.

{ 4 }

COLOR MATERIALS

Flower Classifications

Flowers are classified by life cycle. Annuals, such as cosmos, germinate from seed, grow leaves, flower, set seed, and die within a year. Biennials, such as foxgloves, have a two-year life cycle. During the first year, they germinate and grow leaves, but they don't bloom until the second year of growth. Perennials, such as Lenten rose, are plants that live for more than two years. Bulbs include plants with specialized roots for storing energy and nutrients. They include bulbs, corms, tubers, rhizomes, and tuberous roots.

It's helpful to understand the life cycle of the plants you're considering for your garden project before you put pen to paper. For example, if you love the idea of using kale, lettuce, and annual flowers together in a tiny kitchen garden, you'll need to consider how long each plant type will be used in the garden space. Lettuce and kale, which thrive in the cool temperatures of early spring and late fall, will be harvested after several weeks of growth, leaving room for annual flowers, such as cosmos (in summer) or pansies (in winter). Once you understand the life cycle of your plants, you can extend the season of interest with companion plants. It might be thyme used as an edging, or summer bulbs, such as allium, that can simply be removed once they've flowered, to make room for something new.

Figure 4.1. Shasta daisies, garden phlox, and loosestrife bring a relaxed vibe to the garden.

Figure 4.2. Pansies brighten the winter landscape.

Figure 4.3. Black-eyed Susans and angelonias create a simple display that stands up to summer heat.

Choosing Your Color Palette

Garden designers choose color combinations based on many factors, such as customer preferences, light conditions, and season of interest. One design might call for strong, bold colors for a grouping of pool containers, while another might include cool and sophisticated tones for an entertaining space. Choose color tones and plant types that feel right for *your* outdoor space. If your project includes a planting in your backyard, and you have rambunctious children and dogs, you probably will not choose delicate foliage plants in cool, elegant tones. Instead, you're likely to choose playful colors and forgiving plant material.

Figure 4.4. Angelonias and vincas thrive in sunny gardens and are drought tolerant, once established.

Figure 4.5. Hybrid coneflowers in rich, saturated tones are useful in many types of color displays and bring pollinators into the garden.

Before you consider flowers and other plants for your project, it's helpful to know a little about color theory. If you've ever taken an art class, you'll remember using the color wheel, a simple tool which shows the relationship between colors. On the color wheel, red, blue, and yellow are called primary colors, because they are the foundation of all other colors.

Complementary colors are red/green, violet/yellow, and blue/orange. These colors are located on opposite sides of the color wheel. The classic summer flower bed of red dragon wing begonias is pleasing to the eye because of the natural harmony between the vibrant red blossoms and rich green foliage. Blue salvia paired with orange zinnias is another example of a harmonious combination using complementary colors.

Analogous colors are red/orange, orange/yellow, yellow/green, green/blue, blue/violet, and violet/red. A bed of orange and yellow pansies, or violet and red impatiens, are examples of harmonious combinations using analogous colors. You can also use trios of colors in your designs to create harmony, such as red/orange/yellow.

Another way to create harmony in your flower displays is to use a monochromatic color scheme, such as winter violas in tones ranging from icy blue to deep midnight blue.

All white flower designs add elegance to the landscape, and they work especially well in spaces used in the evening for entertaining. White flowers glow in both moonlight and candlelight, creating a magical setting. In sunlight, white adds a spark to vivid flower combinations.

As you choose a color palette for your project, consider the backdrop for your flowers—your home. If your house is painted in natural tones of gray, tan, brown, or earthy green, you have the most flexibility with color choices. For a home painted white or cream, choose tones such as blue, apricot, golden yellow, rosy pink and spring green, which will show up well against the pale backdrop. Brick homes with red tones look best with flowers in rich color tones. If you have trim painted in a bold color, such as deep purple, create harmony by repeating this color in your flower planting.

Locating Plant Material

Depending on where you live, you may have access to nurseries and garden centers where you can purchase a wide variety of annual flowers, perennials, and other types of plant material. As you plan your garden project, it is helpful to know if the plants you've chosen will be available at your retailer of choice. If you are like most gardeners, you typically visit a garden center and choose from what is available, perhaps having certain plants in mind but open to whatever might turn your head. This plan works well if you only need a few plants, but if you want to purchase plants for a larger project, you might try a different strategy.

Let's say, for example, you need twenty plants of 'Cora Strawberry' vinca for your project. Your local garden center may or may not have these plants on hand when you're ready to do your project. Contact a store manager and ask if they can order *flats* of plants for you. A flat may contain a total of eighteen pots, and this is called an 1801 flat. If it contains fifteen pots, it is a 1501 flat. Annuals may also be sold in quarts or gallons depending on the plant type.

If your local garden center cannot order plants specifically for your project, and they don't have the quantity you need in stock, you have another option. You can search out wholesale plant growers in your area who sell to the public. These companies supply plants in large quantities to landscaping companies and garden designers, and while they typically don't advertise, they sometimes sell to the public. Unlike retail garden centers, wholesale growers are not set up to answer questions and provide guidance, so do your homework before you go. Some wholesalers will post information on plant availability on the Web. These companies can be an invaluable resource and are worth seeking out.

Calculating How Many Plants You Will Need

For containers, you will need approximately four plants (in four-inch pots) of annual flowers per square foot of area for the warm season and five

Figure 4.6. Angelonias
and lantanas are a good
combination for open,
sunny garden spaces.

plants per square foot for the cool season. Cool-season plants will spread less in the mild temperatures of fall, so they are typically planted closer together. Warm-season plants expand in the heat of summer, so using four plants per square foot provides room for growth. You can "eyeball it" to come up with a plant count, or if you like precision, you can calculate the area of your container by measuring the diameter with a tape measure and doing a little math. Once you know the diameter, calculate the area.

Area = 3.14 × the radius of the container (one-half the diameter), squared. For example, if you have a twenty-two-inch container, the radius

Figure 4.7. Dramatic elephant ears, acalyphas, and coleus thrive in the long, hot days of summer.

is eleven inches or 0.92 feet (11/12 = 0.916). Multiply 3.14 × 0.92 × 0.92 and you'll have a total of 2.65 square feet, which you can round off to two and a half square feet. For this example, you'll need about thirteen plants for the cool season and ten plants for the warm season.

For flower beds, measure the length and width of the bed in feet and multiply. Next, subtract out any space that will be taken up by shrubs or large perennials. For example, if your planting includes one shrub approximately two feet by two feet in size, subtract four square feet from the square footage total of your bed.

To calculate how many four-inch pots of flowers you will need, consider how far apart you want to plant them. For the cool season, most landscaping professionals plant pansies and violas at around nine inches on center. As they plant, they measure nine inches from *the center of one plant to the center of the next plant*. Next, they line up the plants in diagonal rows before planting them. If you learn this technique, your beds will look tidy and consistent.

For the warm season, most four-inch annual flowers and foliage plants can be spaced at ten inches on center. Wax begonias, for example, can be spaced as closely as eight inches on center, if you wish to have them fill in very quickly. Plants such as whopper begonias, which expand in summer heat, can be spaced twelve inches on center, or even up to fifteen inches on center. (This is one way to stretch your plant dollars.)

Perennials should be spaced according to their ultimate size. Many perennials such as salvia can be spaced at fifteen to eighteen inches on center, but smaller plants such as ajuga can be spaced more closely, at around twelve inches on center.

When planting shrub groupings, annuals or perennials, most gardeners follow the practice of planting in odd numbers. (You may ignore this rule if the planting exceeds eleven plants.) Your displays will look more natural if you plant in odd numbers, in staggered rows or groupings, instead of lining up your plants in straight lines. For example, if you have three pots of fall pansies to plug into an empty spot in your kitchen garden, group them in a triangle shape (spacing them nine inches on center to allow for growth) instead of a straight line.

Once you decide on plant spacing, you can easily calculate how many plants you will need based on the square footage of the bed:

Square footage × 2.25 = number of plants for eight-inch spacing
Square footage × 1.8 = number of plants for nine-inch spacing
Square footage × 1.44 = number of plants for ten-inch spacing
Square footage × 1.00 = number of plants for twelve-inch spacing
Square footage × 0.64 = number of plants for fifteen-inch spacing

Selecting Your Plants for Seasonal Color Projects

There is a vast amount of plant material available to the home gardener today through local and mail-order plant vendors. In table 4.1, I provide a sample list of plants, grouped by color, along with notations about the type (perennial, annual, bulb), light requirements, and notes on attributes and care. I've included many of the tried-and-true plants grown in southern gardens for decades, as well as new varieties to try in your designs.

Table 4.1. A selection of useful plants for color designs

Botanical Name, Common Name, and Plant Type	Light Requirements
WHITE TO IVORY	
Anemone x hybrida 'Honorine Jobert' (Japanese anemone) (P)	part shade
Catharanthus roseus 'Cora White' (vinca) (A)	sun
Cosmos bipinnatus 'Sonata White' (cosmos) (A)	sun
Digitalis purpurea 'Camelot White' (foxglove) (P)	sun
Euphorbia hypericifolia 'Hip Hop' (euphorbia) (A)	sun to part shade
Gaura lindheimeri 'Geyser White' (gaura) (P)	sun
Helleborus x nigersmithii 'Ivory Prince' (Lenten rose) (P)	part shade to shade
Impatiens x hybrida interspecific 'Bounce White' (impatiens) (A)	part shade to shade
Lantana camara 'Lucky White' (lantana) (A)	sun
Narcissus triandrus 'Thalia' (daffodil) (B)	sun
Paeonia lactiflora 'Festiva Maxima' (peony) (P)	sun
*Penstemon digitalis** (smooth penstemon) (P)	sun to part shade
Pentas lanceolata 'Starla White' (pentas) (A)	sun
Viola x wittrockiana 'Color Max Popcorn' (viola) (A)	sun to part shade
Viola x wittrockiana 'Mammoth White Hot' (pansy) (A)	sun

A = Annual   P = Perennial   B = Bulb, rhizome, or tuber
* Indicates a plant native to North America, Central America, Europe, or Africa

Season of Interest	Notes
late summer to fall	Easy to grow, long season of bloom.
spring and summer	Drought tolerant, don't overwater.
summer	Crisp white flowers for cutting, easy to grow from seed.
spring to summer	May be planted in fall for spring bloom.
spring and summer	Foliage texture works well in containers.
summer	Drought tolerant, airy blooms for beds or pots.
late winter to spring	Good for cold weather containers in a protected area.
spring and summer	Mildew-resistant variety.
spring and summer	Drought tolerant.
spring	Good for naturalizing, elegant white blooms.
late spring	Considered one of the best peonies for the South.
summer	Native plant, good for height at back of bed.
spring and summer	Attracts butterflies.
fall to spring	White and yellow bicolor with large flowers.
fall to spring	Pair pure white blooms with white loropetalum.

Table 4.1. A selection of useful plants for color designs (*continued*)

Botanical Name, Common Name, and Plant Type	Light Requirements
PALE YELLOW TO DEEP GOLD	
Asclepias tuberosa 'Hello Yellow' (butterfly weed) (P)	sun
Calibrachoa x hybrida 'Cruze Yellow' (million bells) (A)	sun
Coreopsis grandiflora 'Corey Compact Gold' (coreopsis) (P)	sun
Echinacea x hybrida 'Sunrise' (coneflower) (P)	sun
Lantana camara 'Lucky Lemon Cream' (lantana) (A)	sun
Ligularia x 'Last Dance' (golden ray) (P)	part shade to shade
*Pachystachys lutea** (golden shrimp plant) (A)	sun to part shade
Rudbeckia fulgida 'Goldsturm' (black-eyed Susan) (P)	sun
Solidago x 'Fireworks' (goldenrod) (P)	sun
Zinnia x hybrida 'Profusion Double Gold' (zinnia) (A)	sun
Viola x wittrockiana 'Delta Pure Lemon' (pansy) (A)	sun
APRICOT TO ORANGE	
Calibrachoa x hybrida 'Persimmon' (million bells) (A)	sun
Canna pretoria 'Bengal Tiger' (bengal tiger canna) (B)	sun
Catharanthus roseus 'Cora Apricot' (vinca) (A)	sun
Chrysanthemum x 'Single Apricot Korean' (daisy mum) (P)	sun
Crocosmia x crocosmiiflora 'Tobias' (crocosmia) (B)	sun

A = Annual   P = Perennial   B = Bulb, rhizome, or tuber
* Indicates a plant native to North America, Central America, Europe, or Africa

Season of Interest	Notes
summer	Native cultivar, attracts butterflies.
spring and summer	Bright yellow blooms, good for container plantings.
summer	Golden blooms contrast well with blues and purples.
summer to fall	Cottage garden plant with citron yellow blooms.
spring and summer	Drought tolerant, lemon yellow and white blooms.
summer to fall	Compact variety useful for small gardens.
summer	Unusual-looking tropical, performs well in heat.
summer	Native cultivar, drought tolerant.
fall	Native cultivar, drought tolerant.
spring and summer	Good for cottage-style gardens.
fall to spring	Pairs well with any blue or purple pansies for winter.
spring and summer	Orange blooms add zing to container plantings.
summer	Green/yellow-banded leaves with orange flowers.
spring and summer	Drought tolerant, do not overwater.
fall	Clouds of airy flowers for fall.
summer	Easy to grow, attracts hummingbirds.

Table 4.1. A selection of useful plants for color designs (*continued*)

Botanical Name, Common Name, and Plant Type	Light Requirements
APRICOT TO ORANGE (*continued*)	
Crossandra infundibuliformis 'Orange Marmalade' (crossandra) (A)	part shade to shade
Echinacea x hybrida 'Sundown' (coneflower) (P)	sun
Impatiens hawkeri 'Sunpatiens Compact Electric Orange' (sunpatiens) (A)	sun to part shade
Kniphofia uvaria 'Echo Mango' (everblooming torch lily) (P)	sun
Perlargonium interspecific 'Caliente Orange' (geranium) (A)	sun
Viola x wittrockiana 'Penny Orange' (viola) (A)	sun to part shade
MAGENTA TO DEEP RED	
Alstroemeria psittacina 'Variegata' (Peruvian lily) (P)	shade
Begonia x hybrida 'Dragon Wing Red' (dragon wing begonia) (A)	part shade to shade
Coreopsis verticillata 'Main Street' (threadleaf coreopsis) (P)	sun
Cuphea llavea 'Tiny Mice' (cuphea) (A)	sun
Dianthus gratianopolitanus 'Firewitch' (clove pinks) (P)	sun
Gomphrena globosa 'QIS Red' (gomphrena) (A)	sun
Mandevilla x hybrida 'Sundenia Red' (dipladenia) (A)	sun
Monarda didyma 'Jacob Cline' (beebalm) (P)	sun
Pentas lanceolata 'Lucky Star Lipstick' (pentas) (A)	sun
Verbena x hybrida 'Vivid Red' (verbena) (A)	sun

A = Annual   P = Perennial   B = Bulb, rhizome, or tuber
* Indicates a plant native to North America, Central America, Europe, or Africa

Season of Interest	Notes
summer	Tropical-looking blooms.
summer to fall	Attracts butterflies to the garden.
spring and summer	Sun-tolerant impatiens, needs consistent watering.
summer	A good vertical element in gardens.
spring and summer	Classic summer container plant, must be deadheaded.
fall to spring	Combine with blues and yellows for lively winter beds.

late summer	Unusual-looking red flowers for late summer color.
spring and summer	Good for containers and beds, a southern classic.
summer to fall	Compact mound of delicate foliage.
spring to fall	Tiny tubular red flowers attract hummingbirds.
spring to fall	Evergreen blue-green foliage.
spring and summer	Tall annual with pops of red flowers.
summer	Heat-loving plant for containers, nonstop bloom.
summer	Mildew-resistant variety.
spring and summer	Attracts butterflies.
spring and summer	Good for mixed container plantings.

Table 4.1. A selection of useful plants for color designs (*continued*)

Botanical Name, Common Name, and Plant Type	Light Requirements
PINK TO FUCHSIA	
Anemone x hybrida 'September Charm' (Japanese anemone) (P)	part shade
Angelonia angustifolia 'Serenita Pink Dwarf' (angelonia) (A)	sun
Begonia x hybrida 'Whopper Green Leaf Rose' (whopper begonia) (A)	part shade to shade
Caladium Bicolor 'Carolyn Whorton' (caladium) (B)	sun to shade
*Delosperma cooperi** (hardy ice plant) (P)	sun
Impatiens x hybrida interspecific 'Bounce Pink Flame' (impatiens) (A)	part shade to shade
Lilium orientalis 'Stargazer' (oriental lily) (B)	sun
Petunia x hybrida 'Supertunia Raspberry Blast' (petunia) (A)	sun
Phlox paniculata 'Maiden America' (garden phlox) (P)	sun
Salvia x hybrida 'Skyscraper Pink' (salvia) (P)	sun to part shade
Scaevola aemula 'Scampi Pink' (fan flower) (A)	sun to part shade
Sedum sieboldii 'October Daphne' (sedum) (P)	sun
Solenostemon scutellarioides 'Kong Rose' (coleus) (A)	part shade to shade
Tulipa 'Pink Impression' (tulip) (B)	sun

A = Annual   P = Perennial   B = Bulb, rhizome, or tuber

* Indicates a plant native to North America, Central America, Europe, or Africa

Season of Interest	Notes
late summer to fall	A staple for shade gardens, easy to grow.
spring and summer	Drought tolerant, compact size.
spring and summer	Largest of the garden begonias, use at back of bed.
spring and summer	Sun tolerant, pink, red, and green foliage.
spring to fall	Drought-tolerant sedum with fuchsia flowers.
spring and summer	Disease-resistant variety.
summer	Bright pink and white blooms.
spring and summer	Good in both containers and beds.
summer	Cottage garden staple for back of the bed.
spring and summer	Tall flower spikes.
spring and summer	Good in containers, drought tolerant.
fall	Silver-blue foliage for year-round interest, fall bloom.
spring and summer	Variegated foliage with fuchsia markings.
early spring	Strong stems with good height make this a favorite.

Table 4.1. A selection of useful plants for color designs (*continued*)

Botanical Name, Common Name, and Plant Type	Light Requirements
LAVENDER TO PURPLE	
Allium x hybrida 'Purple Rain' (allium) (B)	sun
Angelonia angustifolia 'Carita Purple' (angelonia) (A)	sun
Aster x hybrida 'Hella Lacy' (New England aster) (P)	sun
Colocasia esculenta 'Black Coral' (elephant ear) (B)	part shade to shade
Heuchera villosa 'Pinot Noir' (coral bells) (P)	part shade to shade
Impatiens hawkeri x hybrida 'Sunpatiens Compact Purple' (sunpatiens) (A)	sun to part shade
Phlox divaricata 'Ozzie's Purple' (woodland phlox) (P)	part shade
Ruellia brittoniana 'Purple Showers' (Mexican petunia) (A or P)	sun
Salvia farinacea 'Cathedral Purple' (salvia) (A)	sun
Strobilanthes dyerianus 'Persian Shield' (Persian shield) (A)	part shade to shade
Torenia fournieri 'Summer Wave Large Violet' (wishbone flower) (A)	part shade
Viola x wittrockiana 'Delta Lavender Shades' (pansy) (A)	sun
Viola x wittrockiana 'Violet Beacon' (viola) (A)	sun to part shade
LIGHT BLUE TO VIOLET BLUE	
Agapanthus praecox orientalis 'Baby Pete' (lily of the Nile) (B)	sun
Amsonia hubrichtii* (Hubricht's amsonia) (P)	sun
Angelonia angustifolia 'Angelwing Dark Blue' (angelonia) (A)	sun
Calibrachoa x hybrida 'Cruze Blue' (million bells) (A)	sun
Muscari botryoides* (grape hyacinth) (B)	sun

A = Annual   P = Perennial   B = Bulb, rhizome, or tuber

* Indicates a plant native to North America, Central America, Europe, or Africa

Season of Interest	Notes
summer	Purple blooms add whimsy to the garden.
spring and summer	Consistent blooms through the heat of summer.
fall	Pinch back mid-season to keep plants more compact.
summer	Dramatic foliage, lift and store for winter.
year-round	Rich purple foliage.
spring and summer	Sun-tolerant hybrids with vibrant blooms.
spring	Native cultivar.
spring and summer	Typically sold as an annual, can be used as a perennial.
spring and summer	Tall spikes of purple, good for height at back of bed.
spring and summer	Shimmering foliage with deep purple and green stripes.
spring and summer	Good for container plantings.
fall to spring	Lavender color variations create texture and interest.
fall to spring	Purple bicolor blooms.
spring and summer	Overwinter on a sunny windowsill.
summer to fall	Native plant with striking yellow foliage in fall.
spring and summer	Drought tolerant, consistent blooming through summer.
spring and summer	Good in sunny container plantings with other brights.
spring	Vibrant blooms, good for naturalizing.

Table 4.1. A selection of useful plants for color designs (*continued*)

Botanical Name, Common Name, and Plant Type	Light Requirements
LIGHT BLUE TO VIOLET BLUE (*continued*)	
Plumbago auriculata* (cape plumbago) (P)	sun
Salvia guaranitica 'Black and Blue' (blue sage) (P)	sun
Verbena x hybrida 'Deep Vivid Blue' (verbena) (A)	sun
Viola x wittrockiana 'Marina' (pansy) (A)	sun
FOLIAGE PLANTS FROM PALE CHARTREUSE TO DEEP GREEN	
Acoris gramineus 'Ogon' (variegated sweet flag) (P)	part shade
Alpina zerumbet 'Variegata' (shell ginger) (B)	shade
Asarum arifolium* (wild ginger) (B)	part shade to shade
Caladium bicolor 'Miss Muffet' (caladium) (B)	part shade to shade
Carex oshimensis 'Evergold' (variegated sedge) (P)	sun
Cyperus papyrus 'Dwarf Form' (dwarf papyrus) (B)	sun or shade
Duranta erecta 'Gold Edge' (duranta) (A)	sun to part sun
Hakonechloa macra 'Aureola' (hakone grass) (P)	part shade to shade
Hosta x hybrida 'Bright Lights' (variegated hosta) (P)	part shade to shade
Lysimachia nummularia* (creeping jenny) (P)	sun to part shade
Ophiopogon jaburan 'Crystal Falls' (mondo grass) (P)	shade
Sedum rupestre 'Lemon Ball' (sedum) (A or P)	sun
Tiarella wherryi 'Oakleaf' (foamflower) (P)	part shade to shade

A = Annual   P = Perennial   B = Bulb, rhizome, or tuber

* Indicates a plant native to North America, Central America, Europe, or Africa

Season of Interest	Notes
spring and summer	Sky blue flowers, perennial in some zones.
summer to fall	Good salvia for large, sunny beds.
spring and summer	Good for mixed container plantings.
fall to spring	Charming white, deep blue, and light blue flower faces.
year-round	Useful in containers and beds year-round, prefers moist soil.
summer	In most zones it is best to lift and overwinter in a cool place.
year-round	Native plant, good at base of woodland shrub plantings.
spring and summer	Lift and store for the winter, replant in spring.
spring to fall	Grass-like accent plant for beds and containers in sun.
spring and summer	Dramatic foliage, can be overwintered as a houseplant.
spring to fall	Drought tolerant, may be overwintered in a cool place.
spring to fall	Elegant grassy foliage, prefers moist soil.
spring to fall	Lime leaves with deep green edges, good foil for brights.
spring to fall	Versatile plant for containers or used as a groundcover.
spring to fall	Dramatic white wands of flowers in summer.
spring to fall	Bright foliage, planted as an annual in most zones.
spring to fall	Native cultivar, pair with oakleaf hydrangeas.

Table 4.1. A selection of useful plants for color designs (*continued*)

Botanical Name, Common Name, and Plant Type	Light Requirements
WOODY ORNAMENTALS FOR CONTAINERS AND BEDS	
Abelia x grandiflora 'Pinky Bells' (abelia)	sun
Buxus microphylla 'Sprinter' (boxwood)	sun
Camellia japonica 'Winter's Charm' (spring-blooming camellia)	part shade to shade
Camellia sasanqua 'Bonanza' (fall-blooming camellia)	part shade to shade
Chaenomeles speciosa 'Pink Storm' (flowering quince)	sun
Chilopsis linearis 'Burgundy' (desert willow)	sun
Cornus sanguinea 'Artic Sun' (red-twig dogwood)	sun
Distylium 'Blue Cascade' (distylium)	part shade
Edgeworthia chrysantha* (paper bush)	sun
Forsythia x 'Courtasol' (forsythia)	sun
Hydrangea quercifolia 'Ruby Slippers' (oakleaf hydrangea)	part shade
Itea virginica 'Little Henry' (Virginia sweetspire)	sun
Kalmia latifolia 'Elf' (mountain laurel)	sun to part shade
Loropetalum chinense 'Ruby' (Chinese fringe flower)	sun
Loropetalum chinense 'Snow Emerald' (Chinese fringe flower)	part shade
Mahonia 'Soft Caress' (Soft caress mahonia)	part shade to shade
Rhododendron x 'Encore' (encore azalea)	sun to part shade
Vaccinium x 'Sunshine Blue' (blueberry)	sun
Vitex agnus-castus 'Blue Diddley' (vitex)	sun
Weigela florida 'Sonic Bloom' (weigela)	sun

A = Annual   P = Perennial   B = Bulb, rhizome, or tuber

* Indicates a plant native to North America, Central America, Europe, or Africa

Season of Interest	Notes
spring to fall	Pink blooms attract hummingbirds.
year-round	A classic for containers, gives structure to garden spaces.
early spring	Pink, peony-shaped blooms in early spring.
fall to winter	Long season of bloom with compact shape, horizontal form.
spring	Dwarf size is useful in mixed planting beds.
spring and summer	Good for large container plantings, hummingbirds.
year-round	Branches range from yellow to coral red for winter interest.
year-round	Good for container plantings, weeping habit.
late winter	Fragrant flowers on bare stems in late winter.
late winter	Tiny shrub perfect for winter container designs.
early summer	Native cultivar, good fall color.
summer	Pendulous flowers in early summer, native cultivar.
spring	Good for shady areas combined with natives, compact size.
spring and fall	Clusters of pink flowers, rounded shape and compact size.
spring and fall	Fringe-like blooms in spring and fall, compact size.
year-round	Good foliage texture for beds and containers.
spring and fall	Blooms both spring and fall, classic southern garden shrub.
summer	Summer fruit, dwarf size suitable for containers.
summer	Drought tolerant, compact size, dramatic flower spikes.
summer	Good as an anchor for large beds for summer interest.

{ 5 }

HOW TO DESIGN YOUR COLOR GARDEN FOR ALL SEASONS

TEN FLEXIBLE TEMPLATES TO GET YOU STARTED

In my first southern garden, I planted a variety of perennials with no real thought as to how long they would bloom or what my garden might look like through the seasons. Like many gardeners, I simply planted what made me happy. The year started beautifully with daffodils in February, giving way to scillas and tulips. These were followed by spring and early-summer flowering perennials and ornamental shrubs. As late summer approached, I realized that I had put all my eggs in the proverbial garden basket. The lesson I learned was this: there is more to garden design than choosing plants with pretty flowers.

Many years later, while working as an entry-level designer, I was introduced to plant material I had never considered: lacy southern shield ferns, summer-blooming bulbs, quirky vines, tropical plants with patterned foliage, delicate-looking sedums, and bold agaves—it was a new and unexplored world. What I learned was that each plant has a form, a texture, a color. Once you understand the qualities each plant brings to an arrangement, you can put together a tapestry of color and texture that is ever-changing and quite satisfying.

Here, I'm delighted to provide you with ten flexible design templates that can be adapted to fit your own garden. These templates will introduce you to some of the most beloved southern plants, plus new varieties to explore. In my mind, gardening should not be about planting the perfect arrangement of seasonal flowers year after year. It's about experimenting with all of the color nature has to offer and creating something you love.

1. Ode to the Camellia, a Fall Border for Part Shade

I've gardened in areas with varying amounts of shade, from the high shade of mature pines to the deep shade of a woodland path. There are many types of plant material well-suited for morning sun and afternoon shade, and I wanted to create a lively border using plants that bloom happily in part shade. The border should have height, texture, and long-lasting color, with a bit of drama.

I chose two southern favorites as the anchor plants for the design: *Camellia sasanqua* and *Hydrangea paniculata*. Camellias and hydrangeas have been cherished in southern gardens for generations, and I use them often in projects for their beautiful flowers and long season of bloom. I cannot imagine gardening in the South without at least one camellia in my garden. Ten would be better.

Hydrangeas are one of the most popular flowering shrubs in the South, and rightfully so. There are native hydrangeas for dappled woodland plantings and showy cultivars for gardens in full sun. For this design, I've chosen varieties of these two ornamental shrubs, perfect for small garden areas in part shade.

C. sasanqua 'Shishi Gashira' is covered with bright rose-pink blooms from fall to early winter. It is considered a dwarf variety because of its compact growth habit. *C. sasanqua* 'Bonanza' would work equally well here. The evergreen, glossy leaves of the camellia provide an elegant deep-green backdrop for companion plants.

H. paniculata 'Little Lime' is a summer-blooming shrub, with fresh white and chartreuse blooms. This type of hydrangea flowers on new wood, so gardeners typically cut branches back in late winter to twelve inches or so. 'Little Lime' works well here because it can be maintained at a smaller size than other larger hydrangea varieties such as 'Limelight.'

At the base of each camellia, I've used prostrate yew (*Cephalotaxus harringtonia* 'Prostrata'), a low-growing evergreen shrub with a graceful,

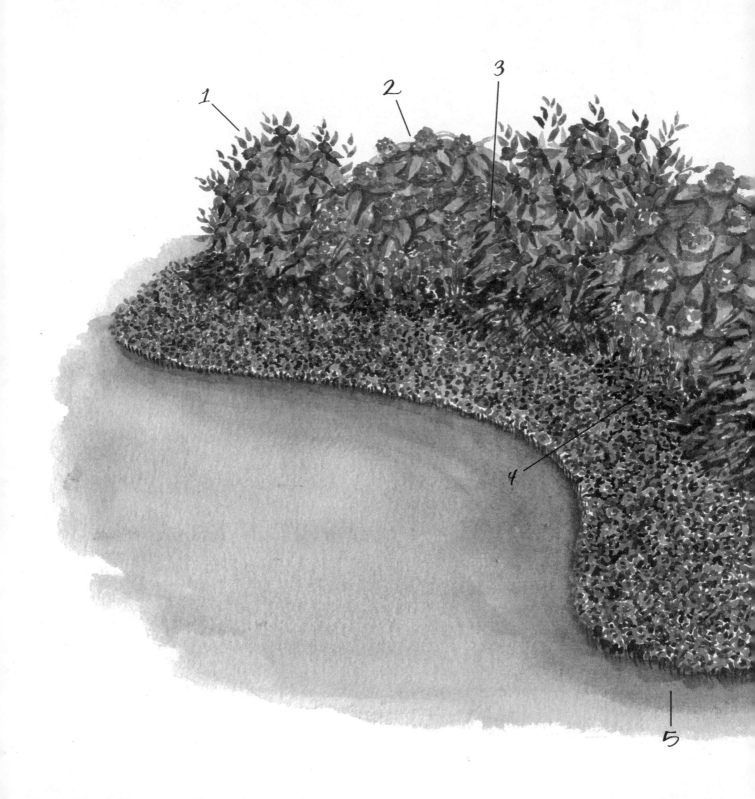

Figure 5.1. Fall-blooming camellias are paired with colorful cool-season annuals, while accent plants hold the promise of spring and summer bloom. Bed dimensions are approximately 6 × 20 feet. In the list below, I've provided spacing information and estimated plant counts.

1 3 (3-gallon) 'Shishi gashira' camellias, spaced 9 feet on center.

2 2 (3-gallon) 'Little Lime' hydrangeas, centered between the camellias.

3 12 (1-quart) purple foxgloves, in groups of 3, planted 12 inches on center, along with 12 'Stargazer' lily bulbs, planted 3 per square foot, at back of bed between shrubs.

4 6 (4-inch) Russian kale, spaced 10 inches on center, along with 6 (4-inch) 'Champagne Bubbles Pink' poppies, planted in front of each kale plant.

5 63 (4-inch) 'Deep Marina' violas, spaced 9 inches on center along the front of the bed.

6 27 (4-inch) 'Light Blue' pansies, mixed with 27 (4-inch) 'Marina' pansies, all spaced 9 inches on center through the back of the bed.

7 3 (3-gallon) prostrate yews, planted closely in front of each camellia.

spreading form and glossy needles. It adds texture and contrast to the arrangement, especially in the winter months. This yew variety is a moderately slow grower and can be pruned a bit from time to time to keep it small. If you wish to do any pruning, be sure to maintain the spreading shape of the shrub.

Why use shrubs instead of perennials and annuals for this design? It's simple. While you could easily design a colorful border using flowering plants with height and interest for the warm season, the cool season is where your design would likely fall short. There simply aren't many cool-season annuals to add height at planting time. A better approach is to anchor the border with shrubs for height and then add annuals, biennials, and bulbs for more color and texture.

The best time to plant shrubs for this mixed border is early fall. Shrub roots continue to grow during the winter in our mild climate and will be less stressed in the summer heat if you follow this practice. Once shrubs are planted and daytime temperatures cool into the seventies, it's time to plant violas, pansies, and accent plants to complete the cool-season color display.

First, plant Russian kale around the base of each hydrangea, where it will add softness during the winter months. This kale selection has glaucous green leaves with fuchsia veining, echoing the rosy pink used elsewhere in the design. 'Lacinato' kale is another beautiful option for the cool season.

Pink Iceland poppies will sit directly in front of the kale, where they will send up a few blooms, then sleep, and bloom again in April. This poppy variety, called 'Champagne Bubbles Pink' (*Papaver nudicaule*), with its crinkled petals and golden stamens, adds whimsy to the garden, and I can't resist adding them to cool-season designs. If your local garden center doesn't carry poppies, contact a commercial grower in your area to check availability, or plant your own from seed in pots in early fall, and transplant into your garden in early spring.

At the back of the bed between the shrubs, plant stargazer lily bulbs (*Lilium orientalis* 'Stargazer') in groups of three. While these showy lilies will not bloom until summer, they are best planted in the fall to allow for root development. (They can also be planted in the spring as soon as soil is workable.)

Once lily bulbs are in place, group purple foxgloves in groups of three or five around them. These tidy green rosettes will expand in spring to give height and drama to the back of the border. Fill the back half of the bed with a combination of pansies 'Pure Light Blue' and 'Maxim Marina,' which has flower faces of deep blue, light blue, and white. Fill the remaining space at the front of the bed with the blue and white bicolored viola 'Deep Marina.' Plant pansies and violas at nine inches on center in staggered rows. When the bed is complete, the flowers should look as if they've been planted on an invisible diagonal grid. This takes some practice, but it gives your flower display a professional look.

While you're enjoying your fall color, do a bit of maintenance to keep it looking its best. As hydrangea blooms fade to brown, snip off spent flowers, leaving the bare branches intact for the next few months. Remove mushy pansy and viola blooms after rainy spells. Use liquid organic fertilizer after harsh cold weather to encourage blooms through the winter. (See chapter 6 for tips on pansy and viola care.)

In late winter, before the hydrangeas begin to show growth, cut each branch to approximately twelve inches. The kale planted in fall around the base of each hydrangea will help to soften the look of the pruned branches before new growth begins in spring. (If you decide to plant another type of hydrangea in your mixed border, such as *H. macrophylla* or *H. quercifolia*, wait to do any pruning until after flowering, as you will forfeit blooms if these types of hydrangeas are cut back in winter.)

If space is limited, and you want to try some of the plants in this design, consider a container planting. Start with a large container measuring between twenty-six and thirty inches in diameter. This will allow plenty of space for the anchor plant to grow until you are ready to change it out for something new. For this simple design, consider using *Camellia sasanqua* 'Kanjiro,' which is a vase-shaped, upright grower with rose-colored blooms in fall. An upright camellia gives height to the arrangement, while allowing space for color around the base of the shrub.

In October or early November, plant a three-gallon camellia in your container, which should be filled with good quality potting soil. Plant Iceland poppies along with Russian kales around the base of the shrub. The poppies will send out a few blooms in fall, rest all winter, and then bloom again in late spring. Kale plants add a ruffle of foliage at the base of the camellia and will help to hide the dormant poppy foliage during the coldest part of the winter. Fill the remaining space in your container with a combination of 'Light Blue' pansies and 'Deep Marina' violas. Add a few plants of vinca 'Illumination,' a trailing vine, to spill over the edge. The evergreen vinca will grow happily in the container year-round.

If you want to add even more color to your container, consider adding spring-blooming bulbs between the plants in late fall. Daffodils, such as the diminutive *Narcissus* 'Tête à Tête,' fragrant hyacinths, or short-stemmed tulips such as 'Angelique' are all good options for container plantings.

2. Dreamy Hydrangeas Fill a Summer Border with Bloom

Hydrangeas take center stage in the summer version of this mixed border, while camellias provide an elegant backdrop for other flowering plants. In early May, it's time to replace pansies with warm-season annuals. Most summer plants prefer warm soil, so don't hurry this transition if your cool-season plants are thriving. Foxgloves and poppies will be blooming well by this month, and if you wish to keep them in the bed until they are spent, simply plant around them.

With the lily bulbs already planted in fall, you can now add 'Aaron' caladium bulbs to the back section of the bed, spacing them about ten inches apart. Caladiums can be purchased in four-inch pots, but in my experience, planting the bulbs yourself produces a more robust plant. 'Aaron,' a classic in southern gardens, is useful because it works in both sun and shade. Keep soil moist, but not wet, and crisp green and white leaves will emerge in several days. In summer, the 'Stargazer' lilies will rise dramatically through the mass of caladiums on tall stems with dark raspberry blooms, edged in white.

If you would prefer to use perennials in place of the caladium bulbs in this design, try *Astilbe x arendsii* 'Deutschland' for lacy white summer blooms. Allow spacing of between fifteen and eighteen inches for these perennial plants.

For the center of the bed, I've chosen dragon wing begonias (*Begonia coccinea* 'Green Leaf Pink'). Lush foliage and prolific blooms make this a southern favorite for beds and containers in part shade. Plant the dragon wing begonias at ten inches on center, if you want them to grow together quickly. If you are willing to wait for the bed to fill in, plant them at twelve inches on center and they will spread nicely through the season. For the final touch, add a few groupings of blue fan flowers (*Scaevola aemula*

Figure 5.2. In early summer, hydrangeas and lilies provide height in this mixed border, while warm-season annuals add layers of bloom. Bed dimensions are approximately 6 × 20 feet. In the list below, I've provided spacing information and estimated plant counts.

1 50 'Aaron' caladium bulbs, spaced 9 or 10 inches apart at the back of the bed.

2 3 (3-gallon) 'Shishi gashira' camellias, spaced 9 feet on center.

3 12 'Stargazer' lily bulbs, planted 3 per square foot between the shrubs.

4 2 (3-gallon) 'Little Lime' hydrangeas, centered between the camellias.

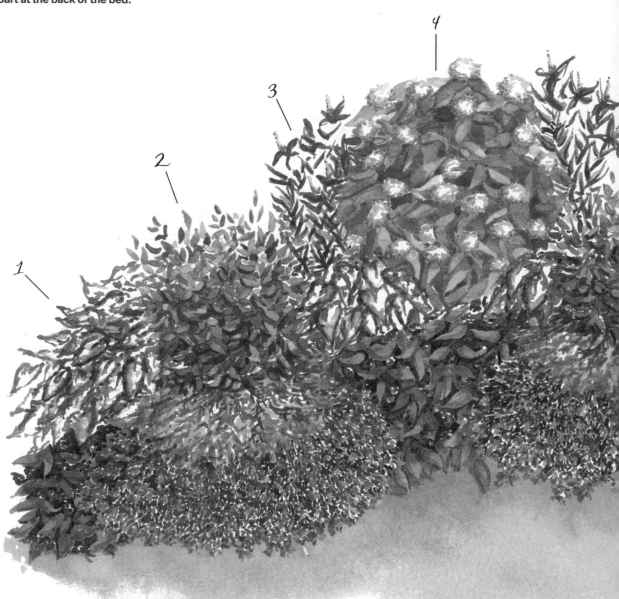

5 27 (4-inch) 'Saphira Blue' fan flowers, spaced
10 inches on center in 3 groups.

6 3 (3-gallon) prostrate yews, planted closely in front
of each camellia.

7 36 (4-inch) 'Green Leaf Pink' dragon wing begonias,
spaced 10 inches on center.

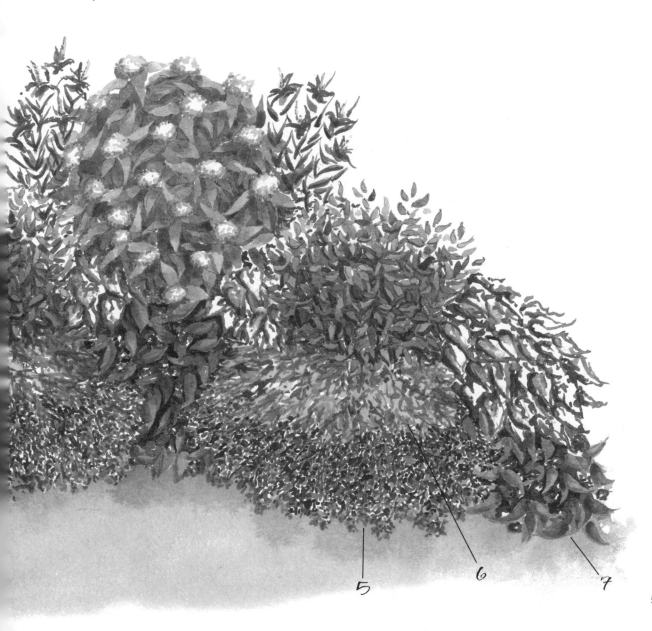

'Saphira Blue'). Fan flowers are heat-loving annuals often used as a low edging for summer flower beds.

I've chosen shades of blue, rose, and white for this design, but other color combinations will work just as well. If you love pops of orange, simply change the poppy color from pink to orange. For summer try mixing equal amounts of pink dragon wing begonias with 'Orange Marmalade' crossandras, which thrive in summer heat. The two plants will mesh together, creating a color combination that is both unusual and pretty. If you want to make plant substitutions, look for alternatives with a similar form and growth habit, whether trailing, low and compact, or tall and open.

To duplicate this design in a container for part shade, use a small camellia as an anchor plant, with a combination of blue wishbone flowers, white fan flowers, and pink dragon wing begonias. Wishbone flower, *Torenia fournieri* 'Summer Wave Large Blue,' is one of my favorite plants for containers, with its charming bell-shaped flowers. The fan flower *Scaevola aemula* 'Saphira White' has delicate white blooms, which are lovely combined with other summer flowers. As the season progresses, the lush foliage of the begonias will weave together with the fan flowers and wishbone flowers and spill over the sides of the container.

3. Abelias and Violas Add a Colorful Twist to a Fall Border

In this design for a sunny border, I wanted to combine a new variety of an old southern favorite, abelia, along with cheerful violas, vibrant dianthus, and colorful bulbs for late winter bloom.

The star of this design is *Abelia x grandiflora* 'Twist of Lime,' with its yellow and green variegated leaves and arching branches. In my grandparents' garden, abelia would have grown to around six feet, but newer varieties are compact, making them easier to use in small garden spaces.

As with other shrub plantings, this one is best planted in the early fall to let roots settle in over the winter. Start with three abelias spaced across the back of the bed. When daytime temperatures fall into the seventies, it's time to plant cool-season annuals. As an accent plant, I've chosen the hybrid dianthus 'Telstar Purple' for its vibrant flowers. Place a few of these at the base of each shrub where they will bloom a bit in fall, rest in winter and then burst into a mass of bloom in spring.

At the back of the bed, plant daylilies (*Hemerocallis* 'Moontraveler') in groups of three between the shrubs. (This reblooming variety has pale lemon blooms from late spring through summer.) As the daylily roots settle in, foliage will eventually be taken down by frost and can be cut to the ground.

To fill the remaining space in the front of the bed, I've chosen a mix of three violas, 'Coconut Swirl,' 'Deep Marina,' and 'Azure Wing,' each one a different variation of blue and white. Space violas at nine inches on center.

In late November or early December, add groups of bulbs directly behind each daylily to boost the color show in early spring. Here I've used a miniature daffodil, 'Jet Fire,' with yellow petals and vibrant orange trumpets. 'Jet Fire' was hybridized from the wild species, *N. cyclamineus*, so it

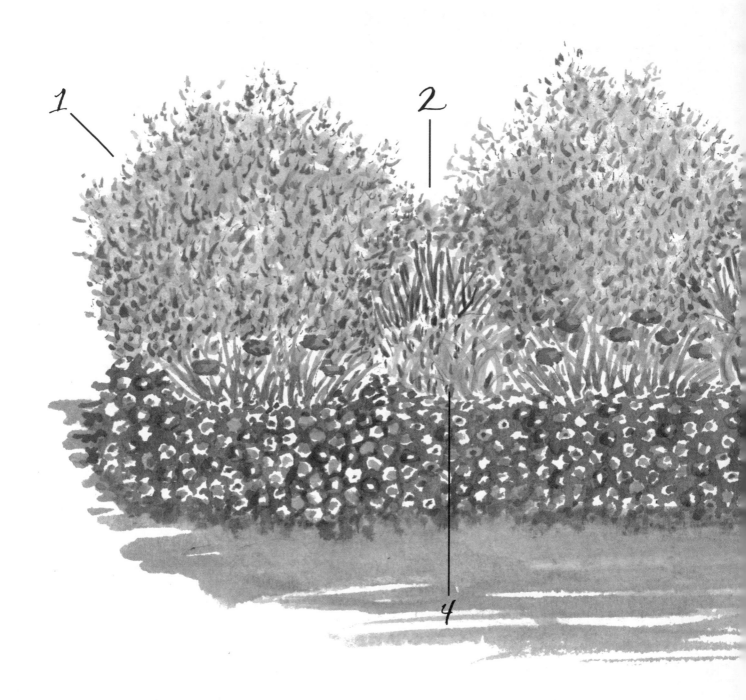

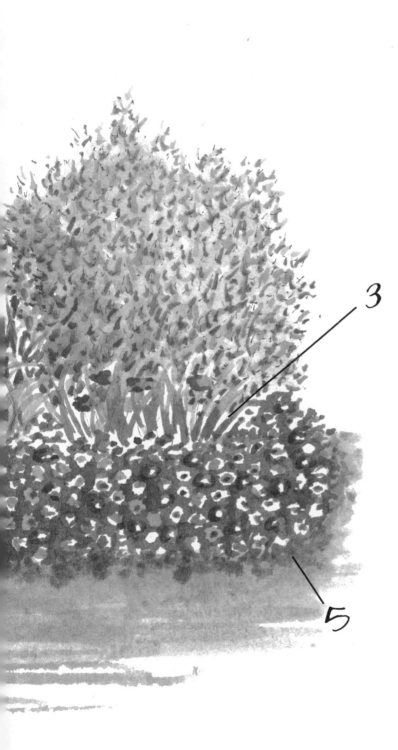

Figure 5.3. In late winter, abelia anchors a bed of daffodils, dianthus, and violas. Bed dimensions are approximately 5 × 12 feet. In the list below, I've provided spacing information and estimated plant counts.

❶ 3 (3-gallon) 'Twist of Lime' abelias, spaced 5 feet on center across the back of the bed.

❷ 50 'Jet Fire' daffodil bulbs, spaced 9 per square foot, in 2 large groups behind daylilies.

❸ 15 (4-inch) 'Telstar Purple' dianthus, planted in 3 groups around the front of the abelias.

❹ 6 (1-gallon) 'Moontraveler' daylilies, planted in 2 groups, 15 inches on center, between the abelias at the back of the bed.

❺ 18 (4-inch) plants each of 'Coconut Swirl,' 'Deep Marina,' and 'Azure Wing' violas, spaced 9 inches on center, to fill the remaining space in the front half of the bed.

does not need to be divided as many modern hybrids do. It will simply spread over the seasons as it would in nature.

In spring, once daffodils have finished blooming, the emerging daylilies will serve as a leafy cover-up, blending with the strappy foliage of the daffodils. The green daffodil leaves must be given several weeks to transfer energy to the bulbs below in order to produce blooms in subsequent years. Once leaves are brown, they can be removed. The technique of pairing daylilies with daffodils has been used for generations and is one to keep up your sleeve.

4. A Sunny Summer Border for Hummingbirds and Butterflies

For the warm-season version of this design, I've chosen tough and forgiving plants that will pair beautifully with the summer-blooming abelia, with its clusters of tiny bell-shaped white flowers. Evergreen leaves mature to a creamy yellow and green, but the intense yellow of its new growth gives this shrub a glowing quality in sunlight. It is related to the honeysuckle, and trumpet-shaped flowers provide a nectar source for hummingbirds and butterflies all summer long.

In late April or early May, remove cool-season annuals, allowing the abelia, dianthus, and daylilies to remain in place. Across the back of the bed, between the shrubs, space 'Serena Blue' angelonias at approximately ten inches on center. *Angelonia angustifolia*, sometimes called summer snapdragon, is a sun-loving plant that will bloom consistently through summer, given regular water, and it will contrast well with the sunny yellow foliage and blooms of adjacent plants. In midsummer, reach into the center of each angelonia and remove entire stems to open its structure and encourage fresh, new growth.

If you would prefer to use perennials as an alternative to the annual angelonia, try *Salvia* x 'Mystic Spires,' for a long season of bloom, from summer into fall. Space these perennials at eighteen inches on center and remove spent bloom spikes through the season.

Through the middle of the bed, space dwarf lantana plants ten inches on center. *Lantana camara* 'Little Lucky Peach Glow' is a beautiful selection, with blooms that open yellow and change to peach as they mature. This lantana variety flowers consistently all summer and will only reach ten to twelve inches in height, giving it an advantage over older varieties.

An edging of fan flower, *Scaevola aemula* 'Scampi White,' completes the warm-season border. This low-growing annual may start out to be a little

Figure 5.4. Summer-blooming abelia attracts butterflies in this garden border with perennials and annuals in shades of yellow, blue, peach, and white. Bed dimensions are approximately 5 × 12 feet. In the list below, I've provided spacing information and estimated plant counts.

❶ 15 (4-inch) 'Serena Blue' angelonias, spaced 10 inches on center through the back of the bed.

❷ 3 (3-gallon) 'Twist of Lime' abelias, spaced 5 feet on center across the back of the bed.

❸ 6 (1-gallon) 'Moontraveler' daylilies, planted in 2 groups, 15 inches on center, planted between the abelias at the back of the bed.

❹ 15 (4-inch) 'Little Lucky Peach Glow' lantanas, spaced 10 inches on center across the middle section of the bed.

❺ 9 (4-inch) 'Scampi White' fan flowers, 10 inches on center in groups of 3 at the front of the bed.

❻ 15 (4-inch) 'Telstar Purple' dianthus, planted in 3 groups around the front of the abelias.

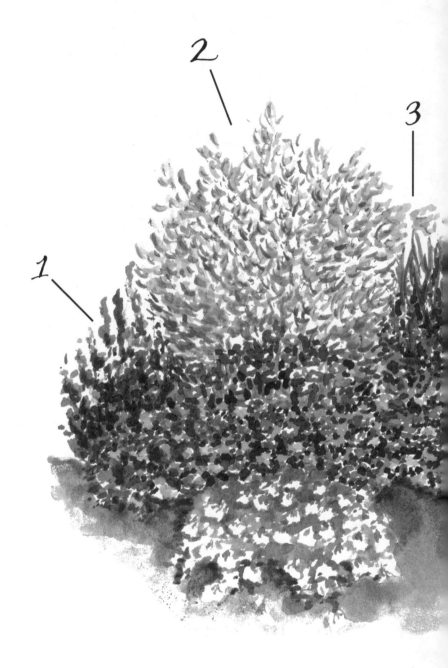

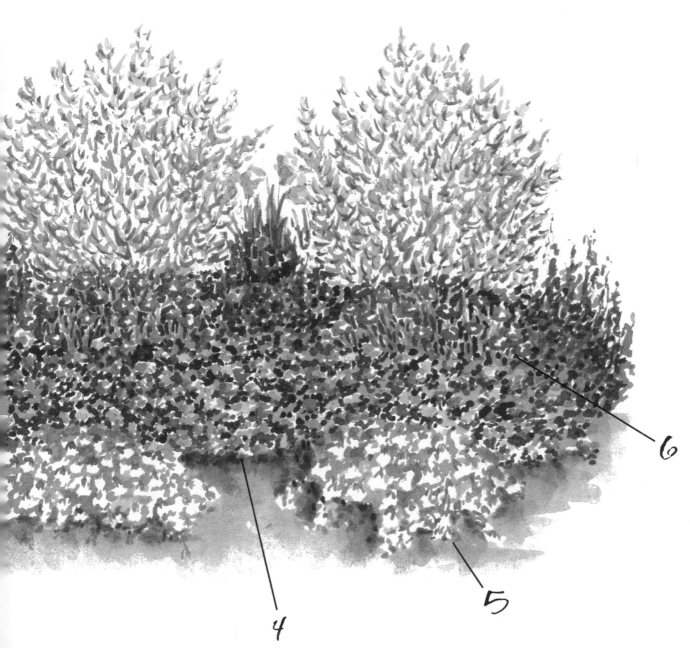

4

5

6

65

wild-looking, but it loves heat and will mesh together beautifully with the lantana by mid-season. By July or August, it may need to be trimmed to keep it in bounds, but fan flowers need no special care, other than regular water.

Give the dianthus a haircut after blooms fade by shearing the top of each plant. You'll soon see fresh growth and new flower buds. If daylily foliage looks tattered by the end of summer, simply cut each plant to the ground and fresh leaves will quickly emerge.

As the abelia matures you can control its size by cutting back the oldest of the whip-like branches to the base of shrub in late winter. This will keep the graceful form of the abelia intact.

5. A Fall Garden Tapestry of Cool-Weather Greens and Edible Flowers

This design is a nod to the colonial kitchen garden, which combined flowers, herbs, fruits, and vegetables to create a rich tapestry of color, texture, and fragrance. These types of gardens were placed close to the kitchen of the home for convenience. The kitchen garden at Colonial Williamsburg in Virginia is a perfect example of how a garden can be both beautiful and functional, and I've found something to admire there no matter what time of year I visit. At home, you can use these same ideas to create a small-scale kitchen garden where you can experiment with edible and ornamental plants and enjoy their simple beauty.

When deciding where to site your kitchen garden, consider building a raised bed along a wall or fence in a sunny back- or side yard. Raised beds are helpful when soil is too heavy or too sandy, because you can amend the native soil with a soil mix specially formulated for these types of beds, creating the perfect environment for both edible and decorative plants.

Planting for this design begins early in the fall season, with three upright rosemary plants anchoring the bed along with two groupings of perennial coneflowers. The fragrant rosemary will be small when you plant it—but have patience. It will grow happily in well-drained soil and will give structure to your garden year-round.

Coneflower or *Echinacea purpurea* is a native wildflower, originally discovered in the 1700s by John Banister, who was studying the native flora of Virginia. New cultivars of this native plant are both colorful and quirky, and for this design I've chosen sunny yellow 'Harvest Moon' to light up the back of the bed from summer into fall.

In the cool-season design, the back half of the bed is filled with a variety of greens and herbs such as Russian kale, lacinato kale, parsley, Swiss chard 'Bright Lights,' chives, lettuce, and spinach. Most greens and herbs can be

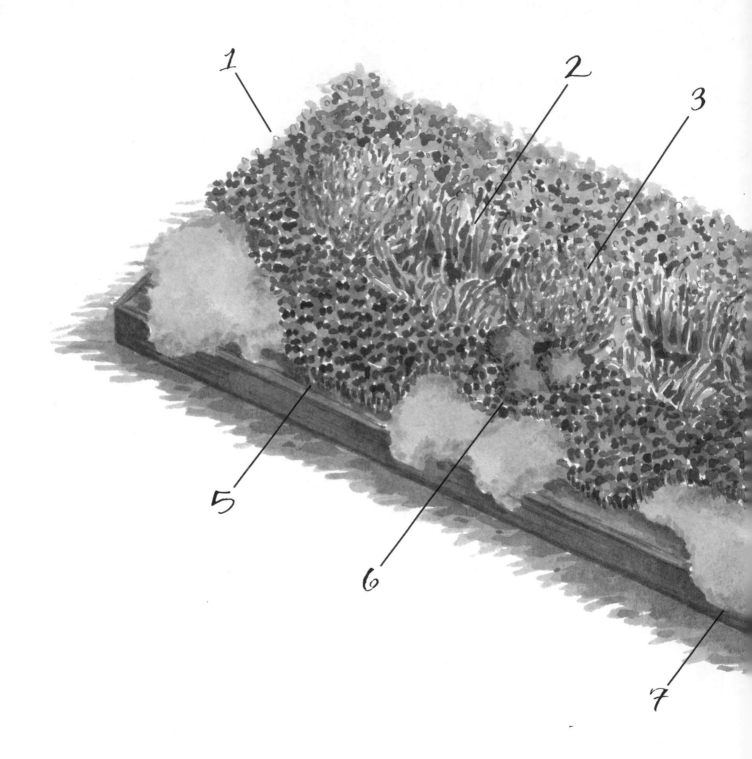

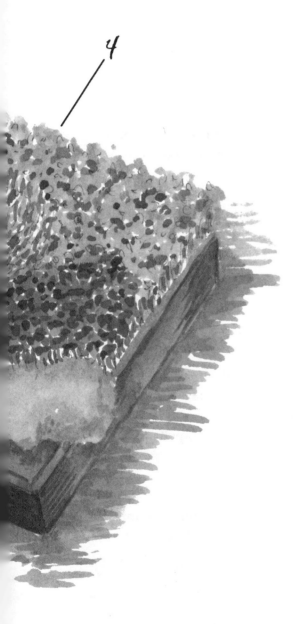

Figure 5.5. In early fall, this kitchen garden combines rosemary with a variety of cool-season greens, herbs, perennials, and annuals for winter interest. Bed dimensions are approximately 5 × 12 feet. In the list below, I've provided spacing information and estimated plant counts.

1 10 Swiss chard plants, spaced 10 inches on center, in 4-inch pots or grown from seed.

2 10 (1-gallon) 'Harvest Moon' coneflowers, planted in 2 groups toward the back of the bed, and 6 (4-inch) chives, planted in 2 groups in front of the coneflower groupings.

3 3 (1- or 3-gallon) rosemary plants, spaced 4 feet on center toward the back of the bed.

4 30 spinach, kale, or lettuce plants, spaced 9 inches on center, in 4-inch pots or grown from seed.

5 18 (4-inch) 'Yellow Jump Up' violas, spaced 9 inches on center across the front of the bed.

6 3 (4-inch) parsley plants, placed in a cluster in front of the center rosemary plant.

7 9 (4-inch) golden thyme plants, spaced 10 inches on center in groups of 3 at the front of the bed, as an edging.

purchased in four-inch pots, and if you would like the bed to look perfect upon planting, simply purchase your herbs, kale, and lettuces in containers for instant gratification.

Another strategy is to plant the back portion of the bed with pots of greens as a decorative element and plant other areas from seed. Planting herbs and greens from seed is easier than you might think. The key is to keep soil moist to allow seeds to germinate. The proper time to plant seed will vary a bit by plant hardiness zone, so do a bit of research, and plant as early as you can according to your specific zone. This will ensure seed germination and a crop to harvest before frost.

Swiss chard 'Bright Lights' has beautiful foliage and bright stems but is not winter hardy, so be sure to harvest these greens, along with lettuce, well before frost visits your garden. Spinach and kale are winter hardy. Parsley, chives, and thyme typically survive our southern winters and can be harvested through the cool season.

To keep the garden looking colorful all winter, plant any open spaces through the middle and front of the bed with 'Yellow Jump Up' violas spaced at nine inches on center. It's a good idea to plant violas by the end of November in most zones to allow them to settle in before winter. This viola has vibrant violet and yellow bicolored edible blooms that can be added to salads. Add three or more pots of golden thyme as an edging to complete the planting. Pinch back violas from time to time to keep them compact and feed with liquid fertilizer to keep them blooming well through the season.

If you want more color in late winter, carefully insert hyacinth bulbs with a bulb planter or trowel in the space between the viola plants in November or early December. In early spring they'll provide fragrance and whimsy to your garden. Once they've bloomed, gently pull up each bulb and allow the violas to fill in as weather warms up.

In the South, many of our gardens include areas shaded out by trees or structures, making it a challenge to grow vegetables. Consider planting a container mini-garden for an area with either dappled sun, or morning sun and afternoon shade. To anchor your container, use the delightful evergreen flowering shrub *Loropetalum chinense* 'Snow Emerald,' which can be paired with a variety of shade-tolerant herbs and greens. 'Snow Emerald' is a dwarf loropetalum with fringe-like white blooms, perfect for shady southern gardens.

To plant this project in early fall, purchase a three-gallon 'Snow Emerald' and place it in a large container with a variety of herbs and greens purchased in pots, such as parsley, chives, beets, lettuce, arugula, kale, or Swiss chard, all of which will grow happily in half a day of sun. Add edible and colorful violas to spill over the edge of the container, where they'll provide beautiful blooms through the cool season.

If you start your project early enough, you can also grow greens from seed directly in the container, keeping the soil moist while they germinate. Check the back of each seed packet for the appropriate month to plant. Many greens, such as spinach, can be planted in late summer for fall harvest.

6. A Spring Kitchen Garden to Delight the Senses

In the spring version of the kitchen garden design, a combination of warm-season vegetables, herbs, and annuals weave together in an explosion of color, texture, and fragrance. Before we delve into the details of the spring and summer season, let's consider the most graceful way to make the transition from cool-season to warm-season plants.

In early spring, cool-season plants will respond to milder temperatures with lush, new growth. When you're ready to do spring planting, harvest any remaining greens, and pull violas to make way for new seeds and plants. The existing coneflowers, rosemary, thyme, chives, and parsley can remain in place.

After the last frost date in your planting zone, plant borage seeds in a staggered row, about ten inches apart at the back of the bed to form a leafy background for shorter plants. Borage is easy to grow and has electric blue, edible flowers. It can simply be pulled in summer when it is done flowering to allow space for the other plants to expand.

In this design I've included a charming dwarf pea plant, perfect for small gardens, which can also be planted after the last frost date. The 'Tom Thumb' pea was one of the hundreds of vegetables grown by Thomas Jefferson at Monticello for its dwarf size and early harvest date. This pea variety is easy to grow from seed and only reaches around nine or ten inches in height, so place it in the front half of the bed where it won't be swallowed by taller plants.

Once seeds have germinated, and weather is warm, fill any remaining space in the back half of the bed with soft herbs such as basil, cilantro, and oregano. While these plants can certainly be grown from seed, I find it satisfying to add herbs in four-inch pots, which give a finished look to the bed.

To complete your kitchen garden display, add annual flowers to the front of the bed. Here I've chosen 'Carita Cascade Deep Purple' angelonia. Vibrant purple flowers keep the bed looking attractive while edible crops are harvested and replaced with new ones. Flowers purchased in four-inch pots give an instant punch of color and contrast to the planting. 'Starla Deep Rose' pentas would be another good option here.

If you would prefer to plant perennials instead of borage at the back of the bed, try *Calamintha nepeta* 'Blue Cloud' for masses of tiny blue flowers all summer.

Heat-loving plants such as peppers can be plugged into empty spots in the garden as greens are harvested. Soft mint, planted in a clay pot and sunk into the garden, is an excellent filler as it is too aggressive to plant as a permanent addition to a small garden space.

If you would like to add fruit to your garden, plant a trio of blueberry plants to replace the rosemary as the structure for the bed. There are small blueberry varieties available that are perfect for smaller gardens. Blueberry production will be most successful if there is a companion blueberry nearby, so if you choose to include them in your garden, be sure to plant at least two.

Vertical space is valuable, and if your vegetable garden happens to be planted against a fence or wall, you can train many types of annual flowering vines and vegetables upward by adding horizontal wire with screw eyes or masonry screws to the structure. Sweet peas, an old-fashioned favorite, are planted in December in my Georgia garden, where they germinate and rest under a bit of pine straw until early spring, when they bolt up any nearby structure. Vegetables such as beans and zucchini can also be grown vertically to save space.

Don't limit your vegetable plantings to a kitchen garden or vegetable garden. Vegetables, herbs, and greens can be added to any sunny planting bed. If you want to edge a flower bed with spinach early in the spring, simply plant seeds and keep them moist until they germinate. Swiss chard planted behind a bed of pansies is both beautiful and edible. Plant cherry tomatoes along with annuals and perennials in a sunny flower bed. Give the tomato plant a support, such as a *tuteur*, and it will mingle happily with the flowers all summer.

Keep your kitchen garden healthy by adding a layer of compost or mulch around the plants to keep weeds out and water your plants well when the soil is dry.

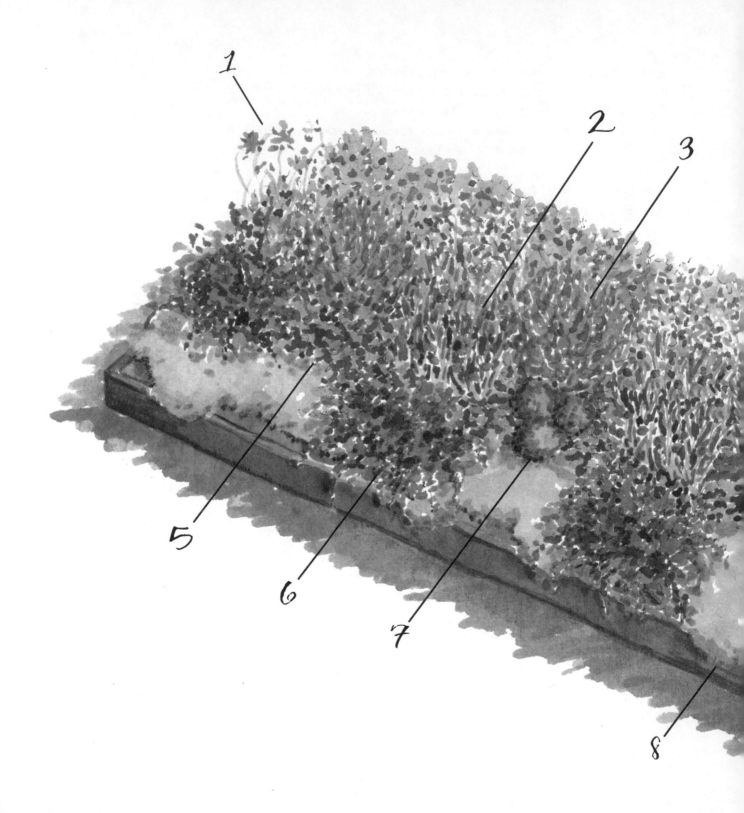

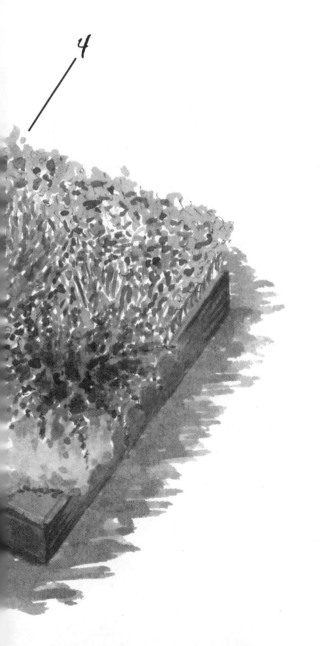

Figure 5.6. In late spring, fragrant rosemary is combined with perennials, annuals, and a variety of herbs and vegetables for a tapestry of color. This warm-season bed is approximately 5 × 12 feet. In the list below, I've provided spacing information and estimated plant counts.

❶ 12 borage plants, spaced 10 inches on center, grown from seed at the back of the bed.

❷ 10 (1-gallon) 'Harvest Moon' coneflowers, planted in 2 groups 12 inches on center, toward the back of the bed, and 6 (4-inch) chive plants, placed in 2 groups in front of the coneflower groupings.

❸ 3 (1- or 3-gallon) rosemary plants, spaced 4 feet on center toward the back of the bed.

❹ 18 soft herbs, such as basil, cilantro, and oregano, planted 10 inches on center, in 4-inch pots or grown from seed.

❺ 4 'Tom Thumb' pea plants, grown from seed, placed toward the front of the bed.

❻ 15 (4-inch) 'Carita Cascade Deep Purple' angelonias, spaced 10 inches on center through the front of the bed.

❼ 3 (4-inch) parsley plants, placed in a cluster in front of the center rosemary plant.

❽ 9 (4-inch) golden thyme plants, spaced 10 inches on center in groups of 3 at the front of the bed, as an edging.

7. A Showy Fall Border with the Promise of Winter Bloom

This fall design is perfect for the gardener who is short on time but longs for interesting color through the seasons. Once planted, this border will look effortlessly beautiful for years to come.

The shrub border begins with *Fothergilla gardenii*, dwarf fothergilla, which grows in moist woodlands from the Carolinas south to Alabama. It has showy, honey-scented, bottlebrush blooms in spring and brilliant fall foliage. Since dwarf fothergilla has showy fall foliage it is typically available in garden centers as weather begins to cool. These shrubs will grow to approximately three feet tall and four feet wide, so space your fothergilla to allow a bit of extra space for bulbs to grow between them for several years. Fothergilla can be shaped slightly to keep it more compact, but little pruning is required.

Flowering bulbs grace the gardens of many older homes in large cities and small towns across the South, and they are an easy way to add color to all kinds of garden spaces. The smallest of the bulbs for this design is *Galanthus elwesii* or giant snowdrop—delightful when planted in groups at the base of shrubs, where they enjoy a bit of shade. The bulbs produce nodding white, bell-shaped flowers in late winter, with foliage reaching four to five inches in height. This variety is thought to be better suited to mild winter climates than other types of snowdrops. If you live in planting zone 8 or 9, try *Leucojum aestivum*, or summer snowflake, an equally beautiful alternative to the giant snowdrop.

Narcissus 'Minnow' is a daffodil hybrid produced from the wild species *N. tazetta*. This fragrant daffodil grows to ten inches tall and has diminutive white and yellow flowers, adding late-winter color to the border. 'Minnow' will thrive and multiply over the years with little care on the part of the gardener.

In late fall or early winter, plant snowdrop bulbs in large groupings at the base of the shrubs. Daffodil bulbs can be planted in groups between the shrubs at the back of the bed and at the outside edges of the border. You'll be adding more types of bulbs to the border in spring, so leave space for the additional bulbs in front of the daffodils.

Your bulbs will multiply slowly over time if left undisturbed, and if you wish to move a few of them, simply wait until they are dormant, when you can dig them up and relocate them. I have been known to move small groups of bulbs just after they've bloomed with a large shovel, keeping the bulbs and soil in a very large clump. Horticulturists would frown at this, but I find that bulbs, along with many other things in the garden, are quite forgiving.

Figure 5.7. Dwarf fothergillas, daffodils, and snowdrops make a charming combination in late winter. This cool-season border is approximately 4 × 12 feet. In the list below, I've provided spacing information and estimated plant counts.

① 50 'Minnow' daffodil bulbs, 5 per square foot, planted in 4 groups, at the back of the bed between the shrubs, and also at the outside edges of the border.

② 3 (3-gallon) dwarf fothergillas, spaced 5 feet on center across the back of the bed.

③ 50 snowdrop bulbs, planted approximately 4 inches apart in groupings in front of each shrub.

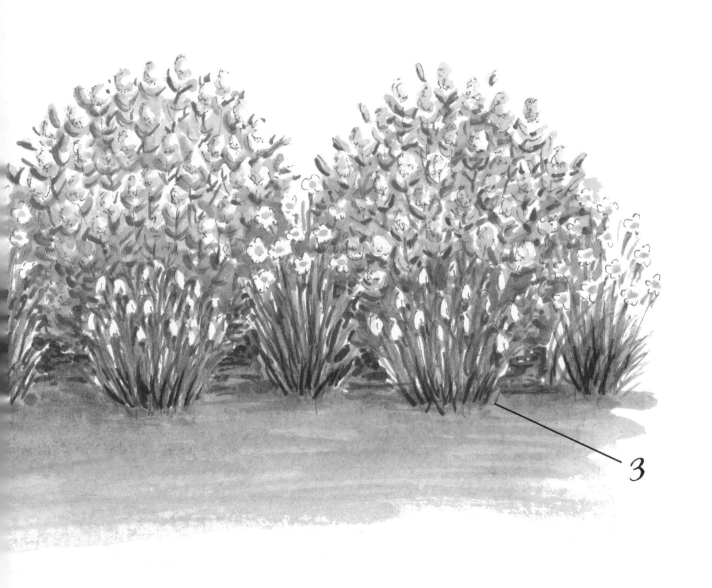

3

8. A Carefree Border with Summer-Blooming Bulbs

This warm-season border, anchored with fothergilla, includes a variety of bulbs for bursts of color through the summer. The first of these is *Zephyranthes* 'Tenexio Apricot,' called rain lilies, because they bloom sporadically through the summer, often after a rain shower. In spring, plant rain lily bulbs in large groupings just in front of the grassy snowdrop foliage, which will persist until it has absorbed enough energy for next year's flowers. (Be sure to allow the snowdrop foliage to turn brown before cutting it to the ground.)

For this design, I've included two types of lycoris for mid and late summer color. The first, *Lycoris squamigera*, is often called resurrection lily. Bold pink, spidery flowers rise dramatically on bare stems in midsummer, followed by strap-like foliage. The coral red, *Lycoris radiata* or spider lily, emerges at the end of summer, just about the time every southern gardener, weary of the heat and humidity, longs for fall color and cooler temperatures.

These bulbs are one of the many types of pass-along plants of our great-grandparents' time, and you'll see evidence of this driving through any small southern town in late summer, where they seem to appear in every front yard. Lycoris bulbs may not bloom the first year they are planted, but you will certainly understand the nickname "surprise lily" when you see these vibrant coral blooms appear in your garden for the first time.

In spring, plant groups of *Lycoris squamigera* bulbs just in front of the green daffodil foliage that will still be visible in the bed (and should be left alone until it turns brown, which means that it has stored enough energy for next year's blooms). Next, plant groups of *Lycoris radiata* bulbs in front of the *L. squamigera* to complete the bulb planting. Your bulbs will root during the spring, and in summer *L. squamigera* will bloom first, sending up strappy leaves to form a fresh green background for *L. radiata*, which will bloom a few months later.

You may preorder both types of lycoris and zephyranthes bulbs from a reputable bulb company in anticipation of spring planting, or purchase them when they're available at your local garden center.

Figure 5.8. Dwarf fothergillas add structure to this planting, with a variety of bulbs for summer and early fall bloom. This warm-season border is approximately 4 × 12 feet. In the list below, I've provided spacing information and estimated plant counts.

1. 40 resurrection lily bulbs (*Lycoris squamigera*), planted 5 per square foot in 4 groups, between each shrub, and also at the outside edges of the border.

2. 3 (3-gallon) dwarf fothergillas, spaced 5 feet on center across the back of the bed.

3. 30 'Tenexio Apricot' rain lily bulbs, planted 4 inches apart in front of snowdrop foliage.

4. 40 surprise lily bulbs (*Lycoris radiata*), planted 5 per square foot in 4 groups, just in front of the resurrection lily bulbs.

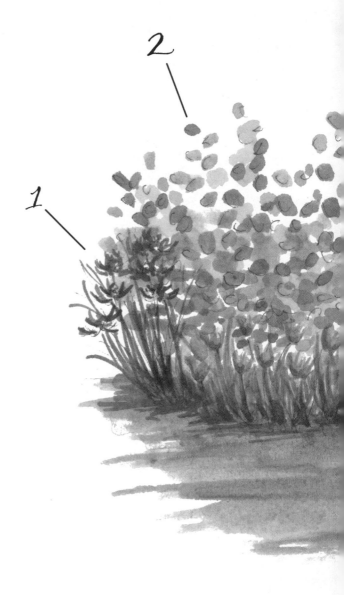

9. Traffic-Stopping Color for Spring and Summer

We're fond of loud colors in the South. Perhaps it has something to do with our hot summers or the clothing we wear to college football games. Maybe it has something to do with the masses of azaleas in eye-popping fuchsia and coral our grandparents and great-grandparents planted, which still explode in bloom across the South every spring. A bit garish perhaps to some—but we love them just the same.

Our annual flower beds should shine just as bright, so I've included a warm-season design with vivid color tones, perfect for a small flower bed along an entrance sidewalk. This design will work equally well for a flower bed in full sun or part shade.

The anchor plant in this design is the tropical *Cordyline fruticose* 'Red Sister,' which has brilliant, striped leaves of fuchsia with a bit of plum. When you plant three cordylines closely together at the back of the bed, they have the visual impact of one large plant, providing plenty of drama as the summer heats up.

Next, plant Coleus 'Wasabi' on either side of the cordylines for contrast. This coleus is one of the more sun-tolerant varieties, with chartreuse, scalloped foliage that is the perfect foil for any bright color.

To continue the neon bright theme, fill the front of the bed with a mix of sunpatiens 'Compact Orange' and 'Compact Deep Rose,' spacing the plants at twelve inches on center to allow for growth. These sun-tolerant annuals, given the botanical name *Impatiens hawkeri*, are hybrids related to *Impatiens walleriana*, which have been beloved in southern gardens for generations because of their vivid colors.

Through the season, keep your coleus compact by pinching off the branch tips every few weeks. The sunpatiens will benefit from light pruning through the season to keep the plants bushy. These plant types need consistent moisture, so water them regularly when there is no natural rainfall. Be careful not to *overwater* in your zeal to help plants grow. If the soil has some moisture in it and the plants look happy, check the bed again in a day or two.

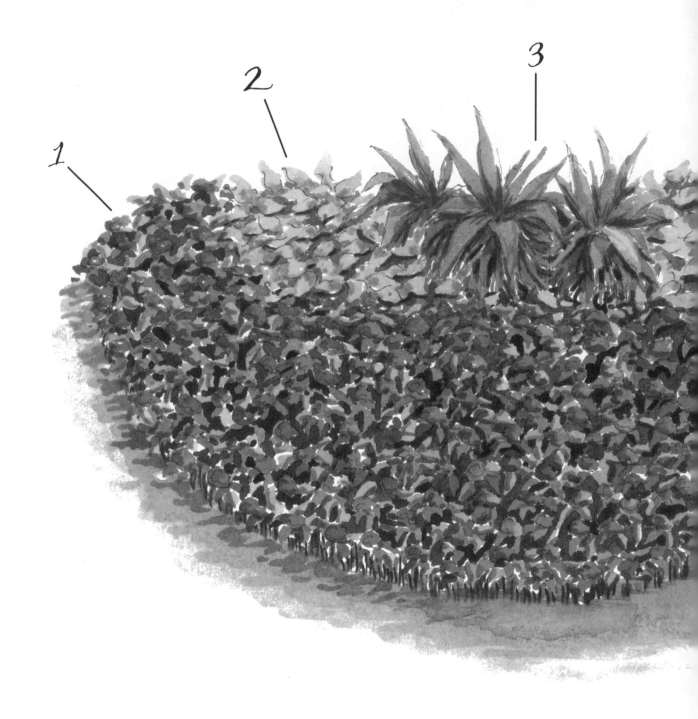

Figure 5.9. Colorful foliage plants are paired with vibrant annuals for non-stop color from late spring through summer. This warm-season bed is approximately 4 × 6 feet. In the list below, I've provided spacing information and estimated plant counts.

❶ 10 (4-inch) plants each of 'Compact Deep Rose' and 'Compact Orange' sunpatiens, spaced 10 inches on center in the front of the bed.

❷ 10 (4-inch) 'Wasabi' coleus, spaced 10 inches on center, in groups of 5 at the back of the bed.

❸ 3 (1-gallon) 'Red Sister' cordylines, planted as one mass, and centered at the back of the bed.

10. Bright Blooms Give Way to Late Winter Fireworks

For the cool season, I've repeated the vivid colors of the warm-season design by pairing the tried-and-true yellow and blue. These primary colors stand out well in the winter landscape and you'll often see variations of the yellow and blue theme in winter garden designs.

Along with the pansies, I've included a brightly colored foliage plant, *Euphorbia x martinii* 'Ascot Rainbow.' Euphorbias are one of the most useful winter accents for southern gardens. The evergreen foliage of this selection is striped with vibrant yellow and green, and it tolerates our winter weather quite well. As a bonus, it sends up attractive, airy sprays of chartreuse green flowers in spring. Plant the euphorbias in groups at the back of the flower bed in odd numbers (three, five, or seven, depending on the size of the bed).

Next, plant blue pansies, spacing them at nine inches on center. Fill the front of the bed with bright yellow pansies to complete the planting.

Tulip bulbs added to the back of the bed provide a jolt of color mid-spring. Here I've combined 'Orange Bowl' and 'Elizabeth Arden' tulips. The key to pairing tulips is to choose selections that share the same bloom season (early, mid, or late), so they will emerge at the same time. When deciding how many bulbs to plant, keep in mind that the rule "less is more" does not apply when it comes to tulips. Plant three to four bulbs per square foot and cover as much of the flower bed as you would like, planting from the back of the bed toward the front. A good rule of thumb is to plant a third of the bed, if budget allows. Simply plant bulbs *between* pansy plants in December when soil is cool, enjoy the bright blooms in spring, and, when blooms fade, *very carefully* remove them to allow the pansies to continue their show. Tulips don't naturalize in the South as they would in colder areas of the country; however, we appreciate the wow factor they give to our winter color displays.

Tip: in the South, tulips perform best with a chilling period of four to six weeks before planting. Order prechilled bulbs from a bulb vendor or refrigerate them yourself, keeping them away from fruits and vegetables.

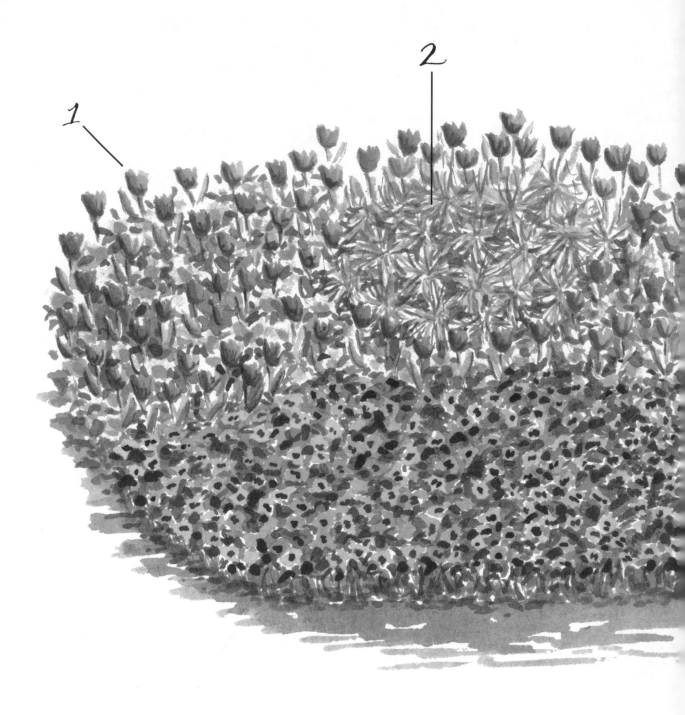

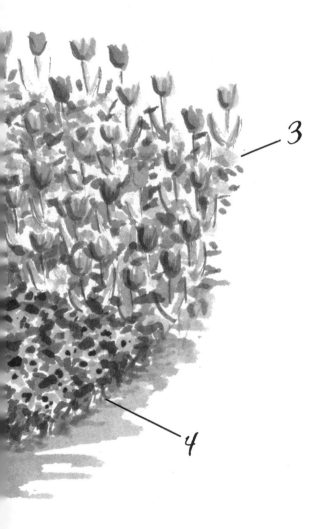

Figure 5.10. This cool-season bed of euphorbias and pansies features a combination of bright pink and orange tulips for a bit of drama in early spring. Bed dimensions are approximately 4 × 6 feet. In the list below, I've provided spacing information and estimated plant counts.

1 30 bulbs each of 'Elizabeth Arden' tulips and 'Orange Bowl' tulips, 4 to 5 bulbs per square foot in the back half of the bed, between the pansy plants.

2 5 (1-quart) 'Ascot Rainbow' euphorbias, planted in a cluster, centered, in the back of the bed.

3 21 (4-inch) 'Panola XP True Blue' pansies, spaced 9 inches across the back of the bed.

4 21 (4-inch) 'Panola XP Yellow' pansies, spaced 9 inches across the front of the bed.

Other Plant Combinations to Explore

Try these plant combinations using woody ornamentals, tropical plants, or perennials for height, along with annuals for added color and interest. Shrubs and perennials can stay in place year after year, while annuals are layered through the bed for color and texture.

1. Combine the finely textured mahonia 'Soft Caress' with 'Whirlwind' variegated hostas and 'Orange Marmalade' crossandras for a sophisticated midsummer display, perfect for a shady area. The airy branches of the mahonia contrast with the wavy foliage of the hostas, creating an attractive backdrop for the crossandras, which love summer heat.

2. If you have a shady garden space with moist soil, try pairing the exotic-looking shell ginger with several plants of variegated sweet flag 'Ogon' for summer. The banded yellow and green leaves of the ginger contrast well with the grassy foliage of the perennial sweet flag. Add violet impatiens for a punch of color. Tropicals, such as shell ginger, can be overwintered in a cool garage or basement and replanted in the garden when soil temperatures are warm in late spring.

3. *Daphne odora* bears sweetly scented flowers in winter, and every southern garden with a shady spot should include one. For color from fall to late spring, plant primrose wallflowers at the base of the shrub and add 'Violet Flare' violas. The pale yellow wallflowers will send up a few blooms in fall, sleep in winter, and bloom profusely in spring along with the violas.

4. The native Piedmont azalea, which has pale pink blooms in late spring, works well in areas with morning sun and afternoon shade. Combine it with 'Ivory Prince' Lenten roses, 'Purple Dragon' lamiums, and 'Sorbet Lavender' violas. To add another season of

interest, add *Clematis crispa*, a native vine, which will climb up through the azalea, adding tiny bell-shaped blooms of blue to the canopy in summer.

5. Try evergreen pieris, or andromeda, to anchor a spring container planting for sun or part shade. Pieris is a southern classic with charming blooms resembling lily of the valley. For the best spring display, plant your container in fall, using pieris 'Interstella,' along with a few 'Telstar Pink' dianthus and the vivid blooms of 'Violet Beacon' violas. The dianthus will bloom a bit in fall, but violas are the workhorse in this arrangement, blooming all winter long.

6. The distylium 'Blue Cascade' is an elegant shrub with arching branches, useful for container plantings in part shade. For the cool season, mix 'White Nancy' lamiums along with 'Color Max Popcorn' and 'Denim Jump Up' violas to spill over the edge of the container. For the warm season, replace the violas with dragon wing begonias or any other shade-tolerant, trailing annuals.

7. Edgeworthia (paper bush) sports pale yellow blooms on bare stems in late winter and is striking when used to anchor a small flower bed or container planting. Combine it with 'Shorty' euphorbias and 'Lavender Blue Shades' pansies to light up a sunny area in winter.

8. For a dramatic fall container planting in sun, combine the curly branches of Corylus 'Contorta' (Harry Lauders' Walking Stick) with 'Bright Lights' Swiss chard, 'Orange Jump Up' violas, orange pansies, and 'Gold Nugget' lamiums.

9. Plant a container with 'Stratosphere White' gaura for height, with 'Carita Cascade Deep Purple' angelonias, pink verbenas, and 'Limoncello' petunias. These sun-loving plants will bloom all summer long.

10. Combine 'Imperial Blue' cape plumbago with pink calibrachoas, white fan flowers, and the quirky yellow blooms of shrimp plant for a colorful poolside container grouping.

11. A combination of pink dipladenia, ice plant, orange portulacas (*Portulaca umbraticola*), and 'Lemon Ball' sedums create a neon-bright container planting for sun that is both showy and drought tolerant. (Tip: be sure to choose a portulaca hybrid such as *Portulaca umbraticola* for summer containers. Moss rose, or *Portulaca grandiflora*, which has needle-like foliage, performs best in northern planting zones.)

12. Anchor a sunny bed with 'Sonic Bloom Pink' weigela to attract butterflies and hummingbirds. Add yellow durantas at the base of the shrub, along with blue fan flowers, for a long season of summer bloom.

13. For lots of color and texture in part shade, 'Lemon Lime' nandina is a perfect partner for the unusual, almost metallic, purple foliage of Persian shield, variegated sweet flag, and violet impatiens.

14. If you're a bird watcher, try native cultivars, such as 'Early Amethyst' beautyberry, 'Whiskers Deep Rose' gaura, and 'Sunrise' coneflower. 'Early Amethyst' is stunning in fall when the branches are lined with jewel-like berries, which birds adore.

Figure 5.11. Garden phlox, anise hyssops, pentas, and fan flowers create a vibrant color display.

{ 6 }

KEEPING YOUR GARDEN LOOKING ITS BEST THROUGH THE SEASONS

You've prepped, dug, and planted, and your colorful flower display is thriving. Now what? Common sense tells you to water plants regularly, but how much water is too much? When should plants be fertilized? What about insect damage? Taking care of garden plants might seem a bit complicated, but it doesn't have to be. Keep it simple and let common sense prevail.

Figure 6.1. Tulips glow in a bed of violas in early spring.

Watering

All plants need adequate water to thrive. But how do you know if you're giving your plants the right amount of water? My horticulture professor liked to say that the correct time to water a plant is a few seconds before it wilts. This statement was always met by some eye-rolling, but his point was this: plants need water only when it is no longer available for them to absorb from the soil. Plant roots need both oxygen and nutrients, and too much water in the soil interferes with the plant's ability to get what it needs from the environment. Plants sitting in waterlogged soil will likely end up on the compost heap.

So how do you really know when it's time to water? Give your plants water when the soil has dried out a bit, but before plants are badly wilted. The simplest way to find out if the soil is dry is to dig down a few inches into the planting bed. If soil still feels a bit moist, wait to water. Water when soil feels dry, preferably in the morning, so foliage has a chance to dry off before evening. Proper watering will ward off most fungal disease issues over the growing season. Rainfall will typically supply a good amount of moisture in the spring most years, but the need for supplemental water will depend on the temperature, soil, and light conditions in your own garden. Shady areas will not dry out as quickly as those in full sun. Sandy soils will not hold on to water as well as clay soils.

Containers will dry out quicker than flower beds, so check them more often during periods of warm weather. A moisture meter is a useful tool to have on hand for this purpose if you have several groups of containers to check.

Monitoring water needs in summer can be tricky, especially during periods of hot dry weather, when plants may wilt a bit in the afternoon sun. Check to see if there is moisture in the soil. If soil still contains some moisture, the plants may simply be reacting to heat, and, if so, they will recover in cooler nighttime temperatures and look refreshed in the morning. For summer containers, try adding water-retaining granules, which help the potting mix hold on to water for longer periods of time.

From fall through winter and into early spring, pansy and viola plantings in beds need to dry out completely before they are watered. Overwatering and water-logged soil can cause disease, especially in the fall when plants are young, so keep this is mind when caring for your cool-season color displays. In the South, winter rains tend to keep soils moist from fall through winter, but if there is a warm, dry spell, check your beds and water them if the soil has completely dried out. This is especially important with container plantings, which can suffer if a period of warm weather is followed by a cold snap. If pansies and violas are sitting in very dry potting mix, they may not be able to cope with a drastic drop in temperature, and the flowers may collapse in the cold.

Topdressing with Compost

Organic gardeners strive to build healthy garden soil containing plenty of organic matter that will decompose slowly, providing nutrients to their plants. I live in an area with plenty of pine and hardwood trees, and when I'm digging there, the soil smells pleasant and woodsy. The forest floor holds layers of leaves, bark, and pine straw in various stages of decomposition, forming a natural mulch over the crumbly clay soil. It is a pleasure to plant in this soil, because nature has enriched it over many years with fallen leaves and tree branches. In my own garden, I want to create this same crumbly soil, rich in organic matter, and topped with some type of organic mulch.

One way to keep your garden soil healthy is by amending it at planting time and then adding compost, when needed, to replenish nutrients. Healthy soil contains millions of microorganisms, which do the important work of breaking down organic matter and making it available to plants. Topdressing flower beds with an inch or two of compost each year will feed the soil while suppressing weed growth.

If you want to learn how to produce compost using materials such as kitchen waste, garden trimmings, newspaper, and leaves, you don't have

to look far. A Web search provides many resources for the home gardener. The Rodale Institute, a nonprofit organization founded in 1947 to support research on organic farming practices, publishes a concise primer on composting basics on its website.

If you don't wish to produce your own, buy the best compost that you can find. If you are buying compost in bulk, be sure what you're purchasing is finely textured, with no chunks of bark or wood, which is an indication it may not be completely broken down. If it smells woodsy or neutral it's likely a very good product. Some composts may have a bit of a manure odor, and this is okay if the texture is good.

Organic Fertilizer

When our great-grandparents grew flowers, they depended on garden compost and composted manure to add nutrients to their garden beds. Many gardeners have returned to this natural approach, which eliminates synthetic fertilizers and instead focuses on building healthy soil so that it can, in turn, provide plants with the nitrogen and other nutrients needed for growth.

Here in the South, organic matter breaks down quicker in our mild temperatures, so there are times when we need to replace nutrients to keep plants thriving. One way to do this is to use organic fertilizers such as fish meal, blood meal, kelp meal, worm castings, and rock phosphate. These fertilizers can be purchased separately or as a premixed product for flowers or specific types of plants, such as hydrangeas and camellias, which prefer acid soil. Perennials, especially southern natives, are more forgiving and may require less nitrogen and other nutrients than annual plants and vegetables.

So how often should you fertilize flower beds? There is no perfect answer. Much will depend on your garden soil and its nutrient levels. Test your soil every few years to determine if adjustments are needed, based on the types of plants you wish to grow. New gardens may benefit from

Figure 6.2. Cape plumbagos and lantanas capture the colors of sunlight and sky in this simple arrangement.

Figure 6.3. Beardtongues (penstemons), are a favorite of bees and hummingbirds.

Figure 6.4. Persian shields and New Guinea impatiens create a striking combination in shade.

the application of organic fertilizer while organic matter is being slowly broken down.

Containers are watered often, sometimes daily in the summer, when daytime temperatures can remain in the nineties for weeks on end. Few of us have the time to make our own potting mix for containers, so most of us purchase bagged potting mix or potting soil, which consists of composted bark, peat moss, and various other materials that vary by brand. Mixes containing fertilizer will feed plants for a month or longer, but because containers are watered often, nutrients can be washed away. Depending on what you are growing in your containers, fertilizer will likely be needed at some point to keep plants thriving, especially in summer. (Most herbs prefer to grow in leaner soil than flowers do and will need less fertilizer.) There are many products to choose from, including organic liquid fertilizers and those formulated specifically for flowers, which will contain a mix of organic ingredients to support bloom production. One option is fish emulsion, which you can use as a foliar spray or liquid feed. There are many excellent books (as well as websites) on organic methods of fertilization, some of which I've listed in the resources section at the back of the book.

Fertilizing Pansies and Violas in Winter

We're typically blessed with mild winters in the South, and this allows us to grow colorful pansies and violas from fall into spring. To keep these flowers blooming well through periods of cold weather, use a water-soluble organic fertilizer. If pansies or violas stop blooming during cold rainy weather, wait until fair weather returns, remove mushy blooms and any damaged leaves, and fertilize. They will recover slowly and resume their bloom cycle, helped along by sunny days and mild temperatures.

Figure 6.5. Tulips highlight a bed of violas in early spring.

Pruning

Many years ago, the owner of a large wholesale plant nursery was answering my questions about warm-season annuals I had used in a few of my projects that weren't performing well. He reminded me that annual plants are healthiest when they are *actively* growing. This statement stuck with me. Simply put, the flowering plant focused on growth is less likely to succumb to disease. You can encourage healthy new growth by pruning. So, what exactly is pruning?

Author and garden designer Tracy DiSabato-Aust explains this concept in her excellent book *The Well-Tended Perennial Garden: Planting and Pruning Techniques*: "When we deadhead spent flowers, pinch stems or buds, and cut back leggy plants we are actually pruning." She defines *pinching* as the removal of only the growing tips of plants along with the first set of leaves (one-half to two inches depending on the type of plant) and *cutting back* as the removal of more than two inches of growth.

If you were to simply water your flowers and refrain from pruning during the growing season, you would see some plants steadily growing over time and some overtaking their neighbors, quite possibly looking a little wild before the end of summer. There is certainly nothing wrong with allowing flowers to do their own thing, but I generally like to see each plant have its time to shine in an arrangement, which means a bit of pruning through the season.

If you are new to this process, start small. Early in the season, if you see that one plant looks a bit leggy compared to its neighbors, simply use your fingers, scissors, or pruning shears to remove the tip of the plant just above a set of leaves. The act of pinching or removing the growing tip of a flowering plant (the terminal flowerhead or bud) sends a signal to the lateral buds, telling them to grow, which creates more flowers and a fuller-looking plant.

Many types of annuals used in flower displays, such as vincas, begonias, portulacas(*Portulaca umbraticola*), crossandras, torenias, petunias, and impatiens are self-cleaning. Old blooms are released and fall off as new blooms emerge.

Sometimes very little is required to keep flowers looking good, but most flower displays benefit from a bit of pruning now and then. Get out your pruners when you see a plant growing faster than its neighbors, or one with a few straggly looking stems. You may decide to reduce the height of a group of plants mid-season. If plants are growing larger and faster than you anticipated, perhaps due to generous rainfall, simply cut them back. When you prune, make your cut just above a leaf or set of leaves.

Some flowering annuals, such as geraniums, will tell you when they need to be deadheaded. After each flush of bloom, spent flower stems will shout, "Hey, I need a little help here!" Geraniums look more attractive when they are consistently deadheaded, so get out your pruners if you love this plant in summer containers. Other examples of annuals that benefit from deadheading or pruning are zinnias, salvias, and the larger varieties of lantanas.

Taller annuals, such as angelonias, will appreciate a mid-season pruning of entire branch sections to open the structure of the plant and encourage new growth. Reach down into the plant to find a few branches that have finished blooming and prune them off. Experiment with a single plant if you are unsure how the plant will react to pruning. Observation is sometimes the best teacher.

Deadheading blooms (or entire bloom stalks) of perennials will help to keep a garden display looking attractive through the season. When old blooms are removed, the plant is forced to use its energy for flower production instead of going to seed. Perennials, such as *Dianthus deltoides*, will bloom heavily in spring; if spent blooms are sheared, the plant will continue to bloom lightly until fall. Some perennials, such as daylilies and stokes asters, can be completely cut to the ground if foliage begins to look ratty at the end of summer, and new growth will emerge. If you're growing native plants, such as coneflowers, you may wish to leave seed heads intact in the fall, as they provide important food for songbirds.

In container plantings, pinching and thinning out stems from time to time is helpful to control overly enthusiastic plants and keep the arrangement looking attractive. For example, if you have a container of 'Cora Cascade' vincas, fan flowers, and Swedish ivy, the fan flowers will eventually

Figure 6.6. Caladiums, vincas, petunias, dragon wing begonias, and creeping jenny create a classic summer container for light shade.

Figure 6.7. Dragon wing begonias, sunpatiens, pentas, begonias, and coleus create a tapestry of color.

be swallowed up by the other, more aggressive plants. To even the playing field, simply thin out the vincas and cut back the Swedish ivy to give the fan flowers space to thrive.

Pansies and violas in cool-season plantings will benefit from deadheading. Both pansies and violas can be pinched back to keep them rounded and compact as cold weather sets in. While the foliage of these plants will not suffer during cold weather, blooms will turn to mush after winter rain and need to be removed when the gardener is able to get to the task. When spring weather arrives, prune them when they begin to get leggy.

How do you know exactly when to prune? It is completely up to you, the gardener. After all, only you can decide on a style for your garden. If you are a tidy sort, you will enjoy the process of pruning or pinching your flower display through the season, and you may find this to be a relaxing task after a long day of work. If you are a free spirit and want to keep things loose and free flowing, you may wish to make only a few adjustments here and there during the season.

Grooming Spent Bulbs

Bulbs store all the energy needed for growth and flower production below ground. When leaves emerge, they absorb sunlight and transfer that energy to the bulb for the next season of growth. Snowdrops, daffodils, and surprise lilies are examples of bulbs often seen in southern gardens. You can add a bit of bonemeal (a source of phosphorous) at the bottom of the planting hole as you plant your bulbs, but they generally need very little care. Some need to be divided from time to time, when flowering diminishes. Once your bulbs have bloomed, allow the foliage to complete its cycle of energy absorption. Once leaves have absorbed all the sunlight they need, they will turn brown and can be removed. If you are overzealous in tidying up spent foliage and remove it too early, you may sacrifice blooms for next season.

Pruning Ornamental Shrubs

When it comes to the task of pruning flowering shrubs, I tend to let nature prevail. Trees and shrubs grow beautifully into the forms they are meant to be based on their genetic makeup. There are many small-sized cultivars of popular shrubs that work well in small garden spaces and require little pruning. Pay attention to the mature size listed on the plant tag when purchasing shrubs. Some deciduous shrubs such as panicle hydrangeas, which bloom on new wood, are typically cut back in the winter before new growth begins, but many other types of flowering shrubs need little pruning at all.

I limit my pruning, in most cases, to the occasional wayward branch. If I decide to slightly reduce the size of an ornamental shrub, I'm careful to maintain its natural shape as I work. If you love the formal look of boxwoods, you can shear them frequently to shape them, but they are equally beautiful in their natural form.

Mulch

Mulch applied to your garden helps to retain soil moisture and minimizes weed germination, and over time, it breaks down, adding organic matter to the soil. Any organic matter such as shredded leaves, compost, or bark chips can be used as mulch. Availability will vary depending on your region, so base your choice for mulch on your native landscape. Pine straw mulch looks appropriate in areas where pine forests are part of the natural landscape, but it might look out of place elsewhere.

Disease and Insect Control

I have a strategy when it comes to insect control in my own garden: be proactive and don't kill the good bugs. Ladybugs eat aphids, and if I'm not killing the ladybugs with pesticides, they will help to control any pests

Figure 6.8. Euphorbias and pansies glow in a winter container planting.

that may be hanging around. The whole point of organic gardening is to maintain the balance between living things above and below the soil. In a healthy garden environment, predators arrive to gobble up the pests, but if you've inadvertently killed the predators, they can't help you out. There's more at stake here than you might think. Pesticides applied in your garden may harm beneficial insects, butterflies, honeybees, and other pollinators, which are critical to our long-term food supply.

Research shows that plants under stress are more likely to succumb to insects and disease, so be sure they are getting adequate water when nature isn't providing rainfall. When you choose to fertilize, use organic products and apply compost to your garden areas to boost the soil microorganisms that support healthy plant life. Remove any diseased plant stems or leaves quickly to prevent the spread of disease. Pull weeds when they are young, before they've set seed, so you're not tempted to resort to chemical weed killers to get them under control. Rotate food crops to help control plant disease whenever possible. This simply means that if you grow tomatoes in your garden, don't plant them in the same spot year after year.

When you mix flowers, herbs, and vegetables in your garden beds, you're creating a diverse environment that mimics nature. The southern gardens of our great-grandparents often contained fruit trees, herbs, and flowers, along with the vegetables that sustained families through the winter months. The botanical complexity of these types of gardens is thought to have a huge advantage—fewer pests.

The time will come when insects are getting on your nerves and you want to do something to get rid of whatever is bothering your plants. First, identify the culprit and then look for the least harmful approach. It may be as simple as dipping leaves in soapy water to remove aphids. Many pests, such as Japanese beetles, can simply be removed by hand and dropped into a bucket of soapy water. There is an "ick" factor here, but it takes a few short minutes to remove the bad guys, and if you spray with pesticides you may very well harm the beneficial insects in your garden that you can't see.

There are many useful books and websites created to support organic gardening practices and to help gardeners identify pests and diseases. See the resources section at the end of the book for more information.

Beauty Is in the Eye of the Beholder

Ultimately, the only person who should decide what your garden looks like is *you*. It may be slightly wild and unruly, or very tidy and neat. Find a garden style and maintenance routine to suit your lifestyle and personality. Garden design should be about what *you* want to see and when you want to see it. If you are very busy in the summer and fear you won't be able to keep up with a garden area, plan your most interesting displays in spring, fall, or even winter. There is much to be said for puttering in the garden on a mild winter day, tidying up pansy beds and checking on the progress of spring bulbs as they emerge. It has become one of my favorite seasons in the garden.

{ 7 }
DESIGN DILEMMAS

This chapter will address some of the more puzzling problems one might encounter while planning a garden project. Sometimes when you're planning a project, you hit a roadblock. Gardeners are an optimistic bunch, and we tend to think there's a silver lining to even the prickliest problems.

Varying Light Conditions in the Same Planting Area

Say you've decided to dress up a terrace with symmetrically placed flower beds, but one is in full sun all day and one is shaded by tree branches in the afternoon. You want to use matching plant material in each bed. How do you choose plants flexible enough to thrive in both sun and part shade?

Figure 7.1. Euphorbias send up clouds of delicate blooms in early spring, giving energy to this bed of blue pansies.

Luckily there are flowers suitable for both sun *and* part shade. For the warm season, try dragon wing begonias or whopper begonias along with sun-tolerant caladiums such as 'Aaron' or 'Red Flash.' Sunpatiens planted with an edging of yellow durantas is another good option.

For the cool season, violas are ideal, as they will bloom well even with a bit of shade during the day. Accent plants such as cabbage, kale, parsley, sweet flag, and euphorbia are good options for both sun and shade in the cool season. Small fall-blooming camellias and spring-blooming pieris, both ornamental shrubs, are also great options to anchor cool-season beds, if you wish to have some height in your winter display.

Plants for Dry, Sunny Areas

Weather is unpredictable, so if your flower bed is in a sun-drenched area that dries out quickly, don't dismay. There are a few steps you can take to help your flowers get through several days without you. One option is to add moisture-retention granules that absorb water and release it back to plants as it is needed. This product is typically used in containers, but it can also be used in flower beds.

Another strategy is to plant drought-tolerant annuals and perennials. Some of these include torch lily, ice plant, sedum, gaura, black-eyed Susan, Mexican petunia, vinca, lantana, portulaca, duranta, and fan flower. Combine yellow lantanas with pink portulacas (*Portulaca umbraticola*) and an edging of blue fan flowers for a colorful and drought-tolerant flower display.

Cool-season flowers, such as pansies and violas, are not likely to be affected by dry spells in the winter. If there is no rainfall during a period of unseasonably warm weather and beds dry out, be sure to water them well. When the winter temperature drops quickly into the twenties or lower, container plants are more likely to suffer, because their root systems are less well insulated than those planted in the ground, so if you are concerned, water plants well before you leave town.

Heat-Loving Pool Plantings

Some of the same strategies for drought-tolerant flower beds can be applied to pool container plantings. Containers are a focal point in any pool area, so they must be both beautiful and tough enough to withstand the unrelenting heat of summer. A proper container planting should have height, so consider grasses or drought-tolerant perennials such as gaura as the centerpiece. Fill each container with bright and colorful plants such as trailing lantana, Swedish ivy, sedum, angelonia, vinca, portulaca (*Portulaca umbraticola*), or fan flower.

Coexisting with Deer

Deer damage is a problem in many areas of the South, even in heavily populated areas. Deer might walk past the same plants for months and then one morning decide to nibble on them, so be proactive. Treat your plants regularly with a granular or spray deer repellant and choose flowers they typically do not like to eat. Deer dislike any flower or herb with strongly scented foliage such as lantana, lavender, rosemary, Russian sage, monarda, nepeta, thyme, yarrow, mint, artemisia, anise hyssop, and chrysanthemum. Poppies, iris, foxgloves and Lenten rose are toxic to deer and they will avoid these plants.

One strategy that has worked well for me is to plant ornamental shrubs for seasonal color, along with bulbs and plants with aromatic foliage where deer are present. If I underplant a shrub with the types of foliage deer dislike, they may nibble but generally move on. Many ornamental shrubs are listed as deer resistant, and I've had good luck with many of them: daphne, false indigo, glossy abelia, viburnum, andromeda, butterfly bush, flowering quince, deutzia, pomegranate, sweet box, blue-mist shrub, beauty bush, and oleander.

Unfortunately, during the cool season, deer love to eat pansies and violas, so if you want to grow them, find areas where you can plant these flowers in containers the deer can't reach, such as a porch, or plant them inside fenced areas. Accent plants such as daffodils, tulips, kale, cabbage, euphorbia, and dianthus are typically not bothered by deer. A simple color display in an area frequented by deer might be a spring-blooming *Camellia japonica* planted with a large grouping of miniature daffodils or snowdrops.

Other Garden Pests

Four-legged pests can be exasperating. Tell a fellow gardener your tale of woe and they will nod in sympathy. There's the armadillo who flattened a young peony with his backside while he ate the tender shoots of nearby plants. The clever rabbit who bypassed a fence lined with chicken wire to squeeze through a miniscule space under a gate and devour an entire strawberry patch while the gardener was away for the weekend. And then there's the tiny vole who nibbled on the roots of lilies, making them topple in a neat row, as if someone had cut them for a flower arrangement.

If you think you might have animal pests feeding on plants, look for signs such as plants cut off below ground level and left uneaten (voles) or stems with clean cuts (rabbits). When deer chew on plant material, they tend to leave uneven damage, while rabbits bite off leaves cleanly with their sharp teeth. If branches have been gnawed and stripped of leaves, this is likely rabbit damage. I have been guilty of blaming deer for damage to shrubs in winter when the culprit was likely a very hungry rabbit.

To discourage rabbits from feeding in your garden, purchase a repellant or mix up homemade pepper spray, which generally consists of cayenne or jalapeno peppers, blended up with water and put into a spray bottle with a few drops of oil and dish soap. Blood meal, something our great-grandparents would have used in their gardens to keep rabbits away, is another option. If you have a mixed bed of vegetables and flowers, adding chives to your planting (which are toxic to rabbits) may also help to keep them away.

Voles prefer to tunnel through soft soil. Discourage them by surrounding planting areas with a band of crushed gravel, a few inches deep.

Armadillos can do a great deal of damage to lawns and gardens, so if you see signs that one of these creatures has decided to move in with you, hire a professional to humanely trap and relocate the animal. Look around the base of shrubs for signs of a burrow. Signs of armadillo damage typically include large holes in turf or planting areas.

Full Shade Areas

If you have a shady garden area, take advantage of the many attributes of a color that is sometimes taken for granted—green. In the spectrum of greens, we can use vibrant yellow-green and deep bottle-green, the pale green of unfurling fern fronds, and the brilliant green of new grass. If you walk along a shady path in a botanical garden, you'll be struck by the subtle textures and patterns that give shade plantings their subdued elegance and sophistication. In these gardens, I'm so enthralled with the contrasts in color, texture, and leaf shape, I forget about flowers completely until I turn a corner and walk back into sunlight again. The play of light and shadow in a shade garden can be magical.

You can create a bit of this drama in an area too shady for flowers, by trying this simple design idea using a container and ground cover. Place a large container in the center of the area and plant a bed of deep green mondo grass around it. Plant the container with a shade-tolerant evergreen shrub such as distylium and then fill with any evergreen plants you choose. Choose a variety of evergreen plants such as creeping jenny, ivy, 'Illumination' vinca, pachysandra, or variegated sweet flag. If you want something more colorful for the warm season, choose plants suitable for low-light conditions (found in the houseplant section of your garden center) and insert them between existing plants in the container.

Figure 7.3. Shell ginger and caladiums thrive in areas too shady for flowers.

Porches and Porticos in Deep Shade

Growing plants in areas such as a screened-in porch or portico in deep shade can be challenging, but the protection they provide from harsh cold often creates a microclimate where you can stretch the limits of what will grow in your area. It also provides an opportunity to experiment with quirky and colorful container plants for warm weather that thrive in deep shade and high humidity.

In the winter, plant sweet box (*Sarcococca ruscifolia*), a small evergreen shrub with fragrant flowers in spring, along with trailing *Vinca minor* and Lenten roses, which can be purchased in full bloom at many garden centers in January and February.

In early summer, when you're ready for a new color arrangement, shop for exotic-looking foliage plants for shade. Rex begonias, with their beautifully patterned foliage, are one of my favorite plants for shady porches, but there are many others to consider. The bold yellow-and-green striped shell ginger is another good option. If you have space for a larger display, combine a philodendron with several interesting foliage plants, some upright and some trailing, to soften the edge of the container. Many of these types of plants can be overwintered in a garage, basement, or a cool room during the winter months and returned to the porch once temperatures are warm.

A Place to Stash Extra Plants

What is a gardener to do with extra plant material? As the seasons pass, you may find that you need a place to stash the seedlings of perennials, plants given to you by gardening friends, or plants purchased for future projects. A nursery bed tucked away somewhere on your property is the perfect home for any assorted plant material you may acquire. A small cedar raised bed, purchased online or at a garden center, is ideal because it requires no soil preparation. Simply purchase a soil mix formulated for raised beds, which will ensure good drainage.

Figure 7.4. Spotted dead nettles and violas soften the edge of a container in light shade.

Moving On

It is difficult to leave behind a garden you created and nurtured. If your new home is an apartment with a small balcony or a townhouse with a small concrete patio, your gardening activity will likely be limited to containers. Don't despair. Containers offer a world of possibilities. Think of a container grouping as a portable garden. Small outdoor spaces can be positively crammed full of containers large and small. It is the ultimate solution for the gardener who likes to shuffle plants around as they come into bloom.

RESOURCES

PUBLICATIONS

Armitage, Allan M. *Armitage's Garden Annuals: A Color Encyclopedia.*
Portland, Ore.: Timber, 2004.

———. *Armitage's Garden Perennials.* Portland, Ore.: Timber, 2011.

———. *Armitage's Manual of Annuals, Biennials and Half-Hardy Perennials.*
Portland, Ore.: Timber, 2001.

Balogh, Anne. "The Best Deer-Resistant Plants for Your Garden." *Garden
Design Magazine*, 2019. https://www.gardendesign.com/plants/deer
-resistant.html#annuals.

Bender, Steve, and Felder Rushing. *Passalong Plants.* Chapel Hill: University
of North Carolina Press, 1993.

Bradley, Fern Marshall, Barbara W. Ellis, and Ellen Phillips, eds. *Rodale's
Ultimate Encyclopedia of Organic Gardening.* Emmaus, Penn.: Rodale, 2009.

Dirr, Michael A. "The Rise of Fothergilla." *Nursery Magazine*, January 2018.
https://www.nurserymag.com/article/fothergilla-michael-dirr.

DiSabato-Aust, Tracy. *The Well-Tended Perennial Garden: The Essential Guide to
Planting and Pruning Techniques.* Portland, Ore.: Timber, 2017.

Garrett, J. Howard. *J. Howard Garrett's Organic Manual.* Fort Worth, Tex.:
Summit, 1993.

Greene, Wilhelmina F., and Hugo L. Blomquist. *Flowers of the South: Native
and Exotic.* Chapel Hill: University of North Carolina Press, 2011.

Hawthorne, Linden. *American Horticultural Society Practical Guides: Gardening
in Shade.* New York: DK, 1999.

Kaufman, Kenn, and Eric R. Eaton. *Kaufman Field Guide to Insects of North
America.* Boston: Houghton Mifflin Harcourt, 2007.

Lawrence, Elizabeth. *A Southern Garden.* Chapel Hill: University of North
Carolina Press, 2001.

Mitchell, Henry. *The Essential Earthman.* New York: Houghton Mifflin, 1981.

Nelson, William R. *Planting Design: A Manual of Theory and Practice.* Champaign, Ill.: Stipes, 2004.

Ritchey, Edwin, Josh McGrath, and David Gehring. "Determining Soil Texture by Feel." University of Kentucky Agriculture and Natural Resources Publications, 2015. https://uknowledge.uky.edu/cgi/viewcontent.cgi?article=1139&context=anr_reports.

Roach, Margaret. *A Way to Garden.* Portland, Ore.: Timber, 2019.

Sullivan, Barbara. *Garden Perennials for the Coastal South.* Chapel Hill: University of North Carolina Press, 2003.

Wade, Gary L., and Paul A. Thomas. "Success with Pansies in the Winter Landscape: A Guide for Landscape Professionals." *University of Georgia Extension Bulletin* 1359 (2012). https://secure.caes.uga.edu/extension/publications/files/pdf/B%201359_2.PDF.

Williams, Charlie. "André Michaux, a Biographical Sketch for the Internet." 2002. http://www.michaux.org/michaux.htm.

WEBSITES

American Horticultural Society. www.ahsgardening.org.

Cooperative Extension System by State. www.nifa.usda.gov/cooperative-extension-system.

Fine Gardening Magazine. www.finegardening.com.

Garden Design Magazine. www.gardendesign.com.

Native Plant Database. www.audubon.org/native-plants.

INDEX